Fluid Mechanics and Its Applications

Volume 117

Aims and Scope of the Series

The purpose of this series is to focus on subjects in which fluid mechanics plays a fundamental role.

As well as the more traditional applications of aeronautics, hydraulics, heat and mass transfer etc., books will be published dealing with topics which are currently in a state of rapid development, such as turbulence, suspensions and multiphase fluids, super and hypersonic flows and numerical modeling techniques.

It is a widely held view that it is the interdisciplinary subjects that will receive intense scientific attention, bringing them to the forefront of technological advancement. Fluids have the ability to transport matter and its properties as well as to transmit force, therefore fluid mechanics is a subject that is particularly open to cross fertilization with other sciences and disciplines of engineering. The subject of fluid mechanics will be highly relevant in domains such as chemical, metallurgical, biological and ecological engineering. This series is particularly open to such new multidisciplinary domains.

The median level of presentation is the first year graduate student. Some texts are monographs defining the current state of a field; others are accessible to final year undergraduates; but essentially the emphasis is on readability and clarity.

More information about this series at http://www.springer.com/series/5980

Yurii N. Grigoryev · Igor V. Ershov

Stability and Suppression of Turbulence in Relaxing Molecular Gas Flows

Springer

Yurii N. Grigoryev
Institute of Computational Technologies
Russian Academy of Sciences
Novosibirsk
Russia

Igor V. Ershov
Institute of Computational Technologies
Russian Academy of Sciences
Novosibirsk
Russia

ISSN 0926-5112 ISSN 2215-0056 (electronic)
Fluid Mechanics and Its Applications
ISBN 978-3-319-55359-7 ISBN 978-3-319-55360-3 (eBook)
DOI 10.1007/978-3-319-55360-3

Library of Congress Control Number: 2017934871

Printed on acid-free paper

This Springer imprint is published by Springer Nature
The registered company is Springer International Publishing AG
The registered company address is: Gewerbestrasse 11, 6330 Cham, Switzerland

Preface

This monograph describes the results of systematic investigations of the authors in the field of stability and initial stages of the laminar-turbulent transition in shear flows of thermally nonequilibrium molecular gases. The dissipative effect arising in such flows due to relaxation of internal degrees of freedom of polyatomic molecules has been recently considered as a tool for increasing stability of these flows and for delaying flow turbulization. Linear stability of plane-parallel shear flows of vibrationally excited gases is studied in the monograph in the general statement. Detailed results on linear and nonlinear stability of a plane Couette flow are presented, including analytical estimates and numerical calculations of the critical Reynolds numbers. Nonlinear evolution of large-scale (coherent) vortex structures and the total cycle of the development of the Kelvin–Helmholtz instability in a thermally excited carrier shear flow are considered.

The mathematical model of flows with relaxation at moderate levels of excitation is based on the full Navier–Stokes equations for a viscous heat-conducting gas with allowance for bulk viscosity. The case of a strongly nonequilibrium vibrationally excited gas is described by the full system of equations of two-temperature aerogasdynamics, where relaxation of vibrational modes is simulated by the Landau–Teller equation for vibrational temperature. The monograph will be useful for aerodynamicists, physicists, mathematicians, and students performing research in the field of hydrodynamic stability theory, turbulence, and flow laminarization.

The book contains an Introduction and seven chapters. The modern status of investigations of the influence of relaxation processes on hydrodynamic stability and turbulence suppression is reviewed in the Introduction. In particular experiments on application of the dissipative effect for control of the laminar-turbulent transition in real hypersonic flows are described.

Chapter 1 has an introductory character and provides some auxiliary material to give an idea of notions and results of physical kinetics, kinetic theory, and acoustics of molecular gases, which are used in the book. The main goal of this chapter is to demonstrate the feasibility and adequacy of mathematical models used in the authors research. In particular the evolution of the concept of bulk viscosity in mechanics and kinetic theory of gases is briefly described, because this phenomenon is still disputable in aerodynamics. Qualitative properties of the Landau–Teller relaxation equation for the vibrational mode energy, which plays a key role in subsequent considerations, are discussed. The physical mechanism of dissipation of acoustic waves on the background of the relaxation process in a thermally nonequilibrium molecular gas is described.

Chapter 2 is devoted to investigations of linear stability of plane-parallel flows of an inviscid nonheat-conducting vibrationally excited gas. Some classical results of the theory of linear stability of ideal gas flows, in particular the first and second Rayleigh's theorems and Howard's theorem, are generalized. An equation of the energy balance of disturbances is derived, which shows that vibrational relaxation generates an additional dissipative factor, which enhances flow stability. Calculations of the most unstable inviscid modes with the maximum growth rates in a free shear layer are described. It is shown that enhancement of excitation of vibrational modes leads to reduction of the growth rates of inviscid disturbances.

Chapter 3 describes the results of numerical and analytical studies of linear stability of a supersonic Couette flow of a vibrationally excited gas. Even and odd inviscid modes of disturbances are analyzed as functions of the Mach number, depth of excitation of vibrational levels, and characteristic relaxation time. The general structure of the spectrum of plane perturbations is studied for finite Reynolds numbers. Two most unstable acoustic viscous modes are identified. Results calculated using the constant viscosity model and Sutherland's law are compared. Neutral stability curves are calculated, which show that the dissipative effect of vibrational mode excitation is inherent in both models of viscosity. The relative increase in the critical Reynolds number caused by excitation is approximately 12%.

An asymptotic theory of the neutral stability curve for a supersonic plane Couette flow of a vibrationally excited gas is developed in Chap. 4. The initial mathematical model consists of equations of two-temperature viscous gas dynamics, which are used to derive a spectral problem for a linear system of eighth-order ordinary differential equations. Unified transformations of the system for all shear flows are performed in accordance with the classical scheme. The spectral problem with two boundary conditions, which was not considered previously in available publications, is reduced to an algebraic secular equation with separation into the "inviscid" and "viscous" parts. The properties of the generalized Airy functions are used for asymptotic estimates of "viscous" solutions. The neutral stability curves obtained on the basis of the numerical solution of the secular equation agree well with the previously obtained results of the direct numerical solution of the original spectral problem.

The energy stability theory extended by the authors to the case of compressible flows of a vibrationally excited molecular gas is used in Chap. 5 to study stability of a subsonic Couette flow. In particular a universal approach is proposed for derivation of equations of the energy balance of disturbances for energy functionals that adequately reflect the evolution of the total energy of oscillations for an arbitrary level of thermal excitation. Based on these equations variational problems are posed for determining the critical Reynolds number of the possible beginning of the laminar-turbulent transition. Their asymptotic solutions are obtained in the limit of long-wave disturbances and yield an explicit dependence of Re_{cr} on the bulk viscosity coefficient, Mach number, and vibrational relaxation time.

Neutral stability curves are calculated for arbitrary wavenumbers on the basis of the numerical solution of eigenvalue problems. It is shown that the minimum critical Reynolds numbers in realistic (for diatomic gases) ranges of flow parameters are reached on modes of streamwise disturbances and increase with increasing bulk viscosity coefficient, Mach number, vibrational relaxation time, and degree of excitation of vibrational modes. The results obtained in the study qualitatively confirm the asymptotic estimates for $\mathrm{Re}_{\mathrm{cr}}$.

Chapter 6 contains the results of the numerical study of a model problem for estimating the influence of thermal relaxation on the turbulized flow outside the limits of the laminar-turbulent transition. Nonlinear evolution of a large-scale vortex structure in a plane shear flow of a molecular gas is considered. Such structures are inevitable attributes of the final stage of the laminar-turbulent transition and turbulence generation in plane wakes, mixing layers, and submerged jets. The results of numerical simulations reported in this chapter lead to a conclusion about a noticeable damping effect of thermal relaxation on nonlinear dynamics of disturbances that can be really reached in nozzle flows, underexpanded jets, or shock waves.

Chapter 7 presents the results of numerical simulations of the full cycle of evolution of the Kelvin–Helmholtz instability, which adequately reproduce the local mechanism of turbulization of the free shear flow. The problem is considered both within the frameworks of the Navier–Stokes equations for a moderate level of thermal nonequilibrium and using the full system of equations of two-temperature aerodynamics for a vibrationally excited gas. Plane waves preliminary calculated by numerical solution of appropriate linearized systems of inviscid gas-dynamic equations are used as initial perturbations. The known pattern of the evolution of the “cat’s-eye” large-scale vortex structure typical for the emergence and development of inertial instability is reproduced in detail. The calculated results show that the relative enhancement of dissipation of the kinetic energy of the structure averaged over its lifetime can reach 12–15% owing to the increase of thermal nonequilibrium in ranges realistic for diatomic gases.

The results presented in the book clearly document the reality of the considered dissipative effect and possibility of its use in control of molecular gas flows.

Novosibirsk, Russia — Yurii N. Grigoryev
December 2016 — Igor V. Ershov

Acknowledgements

The authors express their sincere gratitude to the Russian Foundation for Basic Research, which provided support during the entire cycle of investigations, and to the administration and colleagues from the Institute of Computational Technologies of the Siberian Branch of the Russian Academy of Sciences, in frameworks of which research programs the present investigations occured.

We are also grateful to Profs. Kenneth J. Haller and Mrs. E.V. Medvedeva for their kind assistance in preparing the English text of the manuscript. Last but not least, we would like to thank the editors of the Springer Publishing Company for their kind cooperation and attention in the course of collaboration aimed at publishing this book.

Contents

About the Authors

Yurii N. Grigoryev Professor, Doctor in Physics and Mathematics

Place of work: Institute of Computational Technologies SB RAS.

Professional Awards: The Academician Petrov's Prize of Russian National Committee on Theoretical and Applied Mechanics (2014).

Field of scientific interests: hydrodynamic stability and turbulence, kinetic theory of gases, physical and chemical processes, group methods, mathematical modeling, optimization.

e-mail: grigor@ict.nsc.ru

Igor V. Ershov Professor, Doctor in Physics and Mathematics

Place of work: Department of Information Systems and Technologies, Novosibirsk State University of Architecture and Civil Engineering.

Professional Awards: The Academician Petrov's Prize of Russian National Committee on Theoretical and Applied Mechanics (2014).

Field of scientific interests: hydrodynamic stability and turbulence, kinetic theory of gases, physical and chemical processes, mathematical modeling.

e-mail: i_ershov@ngs.ru; ivershov1969@gmail.com
https://sites.google.com/site/ivershov2011eng/

List of Figures

List of Tables

Introduction

Thermal Relaxation and Stability of Molecular Gas Flows

Relaxation phenomena in molecular gases attracted researchers' attention for the first time in the 1930s due to experiments performed by Kneser [1, 2], who detected and studied the effect of anomalous absorption of ultrasound in a molecular gas (carbon dioxide). Leontovich [3, 4] proposed a clear physical interpretation of this effect in terms of bulk viscosity. A paper of Landau and Teller on the same topic [5] was published simultaneously. Using reduction of the level-by-level description of excitation of vibrational modes of a polyatomic molecule, Landau and Teller derived a new fundamental equation of vibrational energy relaxation, which was named after them. Thus was created the basis of the mathematical apparatus for studying relaxation processes in thermally and chemically nonequilibrium gases.

The modern description of the theory of thermochemical relaxation in gases can be found in the monographs of Clarke and McChesney [6] and of Vincenti and Kruger [7]. In particular the problem of absorption of the energy of acoustic perturbations in the thermal relaxation process in molecular gases was described in detail by Herzfeld and Litovitz [8]. The book [9] was devoted to the kinetic theory of heat and mass transfer and energy exchange in multicomponent molecular gases in which the generalized Chapman–Enskog method was used to derive systems of hydrodynamic equations of different levels of closure, including the equations of two-temperature hydrodynamics.

The development of laser technologies stimulated investigations described in the monographs of Losev [10] and Gordiets et al. [11], where the processes of rotational and vibrational relaxation were considered as applied to gas-dynamic laser problems. The problems of stability of vibrationally excited gas flows were considered for the first time in publications, which were discussed in the reviews of Osipov and Uvarov [12, 13]. Namely, one of the examined aspects was amplification of acoustic waves in a channel flow of a viscous nonheat-conducting gas whose nonequilibrium state was maintained by means of permanent pumping of energy to vibrational modes of molecules and heat removal from translational degrees of freedom. The system of equations of two-temperature relaxation gas dynamics was used, where vibrational relaxation was described by the inhomogeneous Landau–Teller equation with a source characterizing the power of vibrational energy pumping. Spatial localization of the energy input typical for gas-flow lasers leads to generation of gradients of thermodynamic variables in the flow, resulting in reflection of acoustic waves. The inhomogeneous region formed in the flow may act as a resonator where perturbations are enhanced [14, 15]. Two models of energy pumping were considered, where the size of the energy input region was either substantially smaller than the relaxation "length" or comparable with the latter. Investigations of linear stability of the gas flow in the working chamber of the gas-dynamic laser showed that an increase in the pumping region width made the gas flow less stable [15]. At the same time, the loss of flow stability in the working chamber of gas-dynamic laser systems does not mean radiation generation failure: if the vortex mode is enhanced, then the heat transfer in the arising turbulent flow becomes more intense, which ensures better conditions for radiation generation [14].

Investigations of the influence of thermal relaxation on stability of typical aerohydrodynamic flows were started in the 1990s. Nerushev and Novopashin [16] considered the influence of bulk viscosity on the laminar-turbulent transition in the flow in a circular tube. Comparative experiments on the laminar-turbulent transition in the Hagen-Poiseuille flow were performed for nitrogen, N_2, and carbon monoxide, CO. These gases are almost identical in terms of their thermodynamic and transport properties. However, available data show that the bulk viscosity of CO calculated on the basis of ultrasound decay data is several times greater than the corresponding value for N_2 calculated in a similar manner. The laminar-turbulent transition event was fixed through the hydraulic resistance crisis. As a result, it was found that the transition Reynolds number Re_t in a "more viscous" gas (CO) is approximately 10% greater than the corresponding value for N_2. However, reliability of the data obtained in [16] was disputable from the very beginning because of the specific arrangement of those experiments and interpretation of their results. Moreover, there are alternative data for bulk viscosity values for these gases [17–21], which were obtained by advanced measurements of relaxation times behind shock waves in shock tubes. It follows from these data that the bulk viscosity values for N_2 and CO are fairly close to each other, and a small difference between these values cannot be considered as the reason for the observed change in Re_t. In contrast to the ultrasound approach, this alternative method eliminates the influence of gas hygroscopicity, which may be responsible for the large differences in data for CO known to be highly hygroscopic. Nerushev and Novopashin [16] were also aware of these data, but made no comments on the clearly visible contradiction.

Bertolotti [22] performed a pioneering theoretical study of the influence of nonequilibrium internal degrees of freedom on the laminar-turbulent transition in a boundary layer flow. He considered the linear stability of a compressible boundary layer on a semiinfinite flat plate for atmospheric flight conditions at an altitude $H =$ 10 km with the Mach number $\mathrm{M} = 4.5$ and stagnation temperature of 1000 K, which corresponds to motion of real objects. There was no dissociation in the near-wall flow in the chosen regime, but internal degrees of freedom of oxygen, O_2 and N_2

molecules, including vibrational ones, were excited to a sufficiently large extent. It was assumed in the calculations that the nonequilibrium state of rotational and vibrational degrees of freedom is created naturally. In particular, it was assumed that the equilibrium distribution of energy over the degrees of freedom of molecules in wind-tunnel modeling of such flight conditions is violated because of acceleration of the air flow in the nozzle up to a required Mach number. In this case the energy of vibrational modes is "frozen" at the stagnation temperature level, and the static (translational) temperature appreciably decreases. In modeling the motion in an undisturbed atmosphere the translational temperature first increases near the stagnation point, and then the energy redistribution occurs due to acceleration of the flow behind the oblique shock wave on the blunted leading edge of the plate. Adiabatic wall conditions were imposed on the flat plate in both cases. The calculations based on the equations of linear stability theory show that the allowance for bulk viscosity produces a minor stabilizing effect, reducing the amplitude of the second instability mode by several percent. Depending on temperature, the ratio of the coefficients of bulk viscosity η_b and dynamic viscosity η varied in the interval typical for air: $\alpha_1 = \eta_b/\eta = 0.6 - 1$.

A much more powerful and unexpectedly destabilizing effect was observed in the case of a significant deviation from the equilibrium state, when the relaxation process is no longer described by the bulk viscosity model. Instead, the model of two-temperature relaxation hydrodynamics was used. The calculations of stability of a compressible boundary layer in a wind-tunnel experiment on a flat plate with a sharp leading edge showed that the amplitude of low-frequency disturbances of the first instability mode is greater than the value estimated under the thermal equilibrium assumption approximately by a factor of 50. Because of the displacement of the upper branch of the neutral stability curve, the domain of instability of the first mode turned out to be significantly extended in the downstream direction. The calculations for a blunted flat plate moving in an undisturbed atmosphere with due allowance for the nonequilibrium state behind the shock wave revealed that the amplitude of the first mode is twice higher than the corresponding value obtained for the equilibrium conditions. Bertolotti [22] explained this result by significant reduction of the static temperature of the flow because of the excess fraction of internal energy remaining in vibrational degrees of freedom after rapid expansion. It was assumed that rotational degrees of freedom instantaneously reach the equilibrium state with translational degrees of freedom, whereas the energy in vibrational degrees of freedom remains frozen within the characteristic time of the flow. Thus, the change in the compressible boundary layer stability in the considered case is not directly induced by the relaxation process.

A series of experimental studies performed for more than a decade by Hornung group [23] in a unique Caltex T5 reflected shock tunnel. This setup provides the formation of a hypervelocity (both high temperature and high Mach number) flow with the total enthalpy of 3–15 MJ/kg. The main object in those studies was a sharp cone used to study the influence of the growth of the total enthalpy H_0 of the incoming hypervelocity flow of air, N_2, and CO_2, and also air/CO_2 and N_2/CO_2

mixtures on the transition Reynolds number. The laminar-turbulent transition point on the cone was located at the place of extreme growth of the heat flux.

Intermediate results of those studies were summarized in a review paper [23]. It should be noted that from the very beginning, the experiments were aimed at studying the possibility of laminar-turbulent transition delay in hypervelocity flows. The following reasons were formulated for the expected effect. It was known that the laminar-turbulent transition in near-wall flows on a cooled wall at high Mach numbers occurs via the second acoustic mode (cf. [22]). The accompanying processes of vibrational relaxation and dissociation intensely suppress acoustic disturbances. Moreover, under conditions of a hypervelocity flow, CO_2 has a sufficiently large acoustic coefficient of absorption in the frequency range of 10^6–10^7 Hz, where the most intense amplification of the second mode is observed. The latter fact was confirmed by special calculations [24].

Unique relaxation properties of CO_2 have been known to researchers since the experiments of Kneser [1, 2]. The linear triatomic molecule has four vibrational modes: symmetric, asymmetric, and twice degenerate bending mode. The last mode is the most energy-consuming one and is excited at sufficiently low temperatures of the flow $T = 959$ K. The influence of dissociation in pure CO_2 becomes noticeable already at temperatures of $T = 1500$ K.

Relaxation processes in a high-enthalpy flow of CO_2 consist of two stages. At the first (nonstationary) stage, vibrational modes are excited to a certain level at which dissociation begins owing to breakdown of bonds in the course of vibrations. Then there follows the second (quasi-stationary) stage at which the temperatures of vibrational modes remain approximately constant, and the energy flux from translational degrees of freedom is mainly spent on dissociation. The temperature of vibrational degrees of freedom lags behind the static temperature, and the relaxation process consists in replenishment of vibrational energy. This process combined with dissociation ensures absorption of the energy of acoustic disturbances, which leads to an increase in the transition Reynolds number.

All these factors were basically confirmed in subsequent experiments. The main results of those experiments can be formulated as follows. It was shown that an increase in the total enthalpy, H_0, leads to a small increase in Re_t^* for air and N_2 and to a significant increase in Re_t^* for pure CO_2. Here the Reynolds number was determined on the basis of parameters related to the so-called reference temperature. This choice made the data universal in terms of the Mach number, ratio of specific heats, and wall temperature. For a fixed value of H_0, the transition Reynolds numbers, Re_t^*, for air and N_2 were found to be 4–5 times smaller than the value of Re_t for CO_2. The laminar-turbulent transition delay was also detected for mixtures of CO_2 with air and N_2. In particular, the value of Re_t^* for the mixture with N_2 was more than doubled in the case with 40% of CO_2. For the mixture with air, a similar increase was obtained already at 14%.

Thus, the main result of this test series was proof of a significant effect of relaxation processes in hypervelocity flows on the laminar-turbulent transition. At the same time, modest capabilities of the using measurement equipment did not allow reconstruction of the detailed flow pattern. The only measure of thermochemical nonequilibrium was value of H_0 of the flow, which includes the energies of translational, rotational, and vibrational motion of the molecules. During the process, the energy was uncontrollably distributed between different internal modes of molecules and was also spent on dissociation. It is not clear which processes (dissociation, recombination, or energy exchange between vibrational and translational degrees of freedom) prevailed behind the oblique shock wave, in the boundary layer near the cold wall, and in the external flow. Thus, it was impossible to obtain any comprehensive idea about the mechanism of laminarization in the examined flow.

This gap was filled to a certain extent by calculations of Johnson et al. [25] performed for typical experimental conditions in the T5 wind tunnel. The flow was described by the equations of two-temperature aerodynamics supplemented with the equations of convection-diffusion of individual species with allowance for recombination and dissociation. A problem of linear stability of a disturbed conical flow was considered within the framework of this model. An important issue was to take into account the contributions of linearization of the dependencies of the transport coefficients, relaxation time, and reaction rate constants on temperatures and concentrations. After detecting the beginning of growth of the most unstable mode, the coordinate of the laminar-turbulent transition end (critical Reynolds number) was calculated with the use of the N-factor. The calculations reproduced the increase in the critical Reynolds number due to the increase in the total enthalpy of the flow and also the greater values of Re_t^* for pure gases and mixtures with lower thresholds of vibrational excitation and dissociation. The calculated values of the Re_t^* for air and N_2 were higher than the corresponding experimental values approximately by a factor of 1.5–2. This difference was attributed to a high noise level of the free stream generated in the T5 wind tunnel, which stimulated an earlier laminar-turbulent transition in experiments. Unfortunately, the majority of the results of [25] were obtained for less interesting flows of air and N_2. A comparative contribution of vibrational relaxation and dissociation (recombination) in different regions of the flow and other important issues of interrelated thermochemical processes were also not reported in sufficient detail.

The experiments of [23] were followed by investigations of the possibility of controlling the laminar-turbulent transition and thermal protection by means of CO_2 injection into the boundary layer on the cone surface. It is known that such elements of thermal protection systems are used in various hypersonic flying vehicles. Those activities were continued with recent experiments performed in the same wind tunnel [26, 27]. The results of these experiments and corresponding computations were summarized by Leyva [27]. The main problem was to heat the injected gas in the hot boundary layer up to temperatures of the order of 2000 K without inducing a premature laminar-turbulent transition directly at the place of injection. Various schemes of injection through holes and pores were considered. The results of experiments and computations turned out to be somewhat contradictory. Thus, all experiments with holes resulted in the becoming a laminar-turbulent transition instead of its delaying. The same results were predicted by linear stability calculations of Wagnild et al. [28] performed using a scheme similar to that of [25], but with the use of commercial software. On the other hand, the calculations reported in [23] and performed at a sufficiently high CFD level showed that the temperature of CO_2 in the boundary layer rapidly reaches 2000 K. This is sufficient for the beginning dissociation and vibrational relaxation, which is expected to increase boundary layer stability. Some experiments with porous injection revealed an approximately 15% increase in the transition Reynolds number, which then drastically decreased as the injection intensity was increased. However, the calculations of Wagnild et al. [28] for porous injection predicted the laminar-turbulent transition immediately after the porous insert.

Comprehensive investigations on supersonic flow control with the use of initiation of relaxation processes are also performed now by other research teams [29–31].

It should be noted that the interest in practical application of this effect in hypersonic flying vehicles continues to increase, though the available experimental data and their theoretical justification are still far from implementation in real structures.

This brief retrospective review shows that there were no systematic results on stability of flows of thermally nonequilibrium molecular gases at the early 2000s, though various applications urgently require such investigations. Our understanding of such a challenge stimulated our activities described in [32–42], which form the basis of this book.

References

1. Kneser, H.O.: Zur Dispersionstheorie des Schalles. Ann. Phys. **403**, 761–776 (1931)
2. Kneser, H.O.: Die Dispersion hochfrequenter Schallwellen in Kohlensäure. Ann. Phys. **403**, 777–801 (1931)
3. Leontovoich, M.A.: Remarks on the theory of sound attenuation in gases [in Russian]. Zh. Éksp. Teor. Fiz. **6**, 561–576 (1936)
4. Leontovoich, M.A.: Some questions in the theory of sound in absorption polyatomic gases [in Russian]. Izv. Akad. Nauk SSSR. Ser. Fizika **5**, 633–636 (1936)
5. Landau, L., Teller, E.: On the theory of sound dispersion. In: Ter Haar, D. (ed.) Collected Papers of L.D. Landau, pp. 147–153. Pergamon Press, Oxford (1965)
6. Clarke, J.F., McChesney., M.: The Dynamics of Real Gases. Butterworths, London (1964)
7. Vincenti, W.G., Kruger, C.H.: Introduction to Physical Gas Dynamics. Krieger, Malabar (1986)
8. Herzfeld, K.F., Litovitz, T.A.: Absorption and Dispersion of Ultrasonic Waves. Academic Press, New York (1959)
9. Nagnibeda, E.A., Kustova, E. V.: Non-equilibrium Reacting Gas Flows. Kinetic Theory of Transport and Relaxation Processes. Springer, Berlin, Heidelberg (2009)
10. Losev, S.A.: Gasdynamic Laser. Springer, Berlin, Heidelberg (1981)
11. Gordiets, B.F., Osipov, A.I., Shelepin, L.A.: Kinetic Processes in Gases and Molecular Lasers. Gordon and Breach, New York (1987)
12. Osipov, A.I., Uvarov, A.V.: Kinetic and gasdynamic processes in nonequilibrium molecular physics. Sov. Phys. Usp. **35**, 903–923 (1992)
13. Osipov, A.I., Uvarov, A.V.: Stability problems in a nonequilibrium gas. Phys. Usp. **39**, 597–608 (1996)

14. Mukin, R.V., Osipov, A.I., Uvarov, A.V.: Interaction of small hydrodynamic perturbations with a nonequilibrium region in a gas flow. J. Appl. Mech. Tech. Phys. **46**, 818–823 (2005)
15. Mukin, R.V., Osipov, A.I., Uvarov, A.V.: Stability of an inhomogeneous vibrationally-nonequilibrium flow in a waveguide. Fluid Dyn. **42**, 126–132 (2007)
16. Nerushev, A., Novopashin, S.: Rotational relaxation and transition to turbulence. Phys. Lett. A **232**, 243–245 (1997)
17. Mikhailov, I.G., Soloviev, V.A., Syrnikov, Yu.P.: Fundamentals of Molecular Acoustics [in Russian]. Nauka, Moscow (1964)
18. Stupochenko, E.V., Losev, S.A., Osipov, A.I.: Relaxation Processes in Shock Waves [in Russian]. Nauka, Moscow (1965)
19. Kaye, G.W., Laby, T.H.: Tables of Physical and Chemical Constants. Longmans, Green & Co., London (1958)
20. Kikoin, I.K. (ed.): Tables of Physical Quantities, Handbook [in Russian]. Atomizdat, Moscow (1976)
21. Grigor'ev, I.S., Meilikhov, E.Z. (eds): Physical Quantities, Handbook [in Russian]. Énergoatomizdat, Moscow (1991)
22. Bertolotti, F.B.: The influence of rotational and vibrational energy relaxation on boundary-layer stability. J. Fluid Mech. **372**, 93–118 (1998)
23. Leyva, I.A., Laurence, S.J., Beierholm, A.W.-K., Hornung, H.G., Wagnild, R.M., Candler, G.V.: Transition delay in hypervelocity boundary layers by means of CO_2/acoustic instability interactions. AIAA Paper 2009-1287 (2009)
24. Fujii, K., Hornung, H.G.: Experimental investigation of high-enthalpy effects on attachment-line boundary layer transition. AIAA Journal **41**, 1282–1291 (2003)
25. Johnson, H.B., Seipp, T., Candler, G.V.: Numerical study of hypersonic reacting boundary layer transition on cones. Phys. Fluids **10**, 2676–2685 (1998)
26. Beierholm, A.K.-W., Leyva, I.A., Laurence, S.J., Jewell, J.S., Hornung, H.G.: Transition delay in a hypervelocity boundary layer using nonequilibrium CO_2 injection. GALCIT Technical Report FM 2008.001, California Institute of Technology, Pasadena (2008)
27. Leyva, I.A.: Transition delay in hypervelocity boundary layers by means of vibrational relaxation and acoustic instability interactions. AFRL Technical Report 2013-0054, Air Force Research Laboratory, Edwards AFB, California (2014)
28. Wagnild, R.M., Candler, G.V., Leyva, I.A., Jewell, J.S., Hornung, H.G.: Carbon dioxide injection for hypervelocity boundary layer stability. AIAA Paper 2010-1244 (2010)
29. Gembarzhevskii, G.V., Generalov, N.A., Solov'ev, N.G.: Investigation of the velocity fluctuation spectrum of a vortex flow of vibrationally excited molecular gas in a glow discharge. Fluid Dyn. **35**, 222–231 (2000)
30. Vinnichenko, N.A., Nikitin, N.V., Uvarov A.V.: Karman vortex street in a vibrationally nonequilibrium gas. Fluid Dyn. **40**, 762–768 (2005)
31. Kirilovskiy, S.V., Maslov, A.A., Poplavskaya, T.V., Tsyryul'nikov, I.S.: Influence of vibrational relaxation on perturbations in a shock layer on a plate. Tech. Phys. **60**, 645–655 (2015)
32. Grigoryev, Yu.N., Ershov, I.V.: Relaxation-induced suppression of vortex disturbances in a molecular gas. J. Appl. Mech. Tech. Phys. **44**, 471–481 (2003)
33. Grigor'ev, Yu.N., Ershov, I.V., Ershova, E.E.: Influence of vibrational relaxation on the pulsation activity in flows of an excited diatomic gas. J. Appl. Mech. Tech. Phys. **45**, 321–327 (2004)
34. Grigor'ev, Yu.N., Ershov, I.V.: Effect of bulk viscosity on Kelvin-Helmholtz instability. J. Appl. Mech. Tech. Phys. **49**, 407–416 (2008)
35. Grigor'ev, Yu.N., Ershov, I.V.: Energy estimate of the critical Reynolds numbers in a compressible Couette flow. Effect of bulk viscosity. J. Appl. Mech. Tech. Phys. **51**, 669–675 (2010)
36. Grigoryev, Yu.N., Yershov, I.V.: The linear stability of inviscid shear flow of vibrationally excited diatomic gas. J. Appl. Math. Mech. **75**, 410–418 (2011)

37. Grigor'ev, Yu.N., Ershov, I.V.: Dissipation of vortex disturbances in a vibrationally nonequilibrium diatomic gas. Thermophys. Aeromech. **19**, 183–192 (2012)
38. Grigor'ev, Yu.N., Ershov, I.V.: Critical Reynolds number of the Couette flow of a vibrationally excited diatomic gas energy approach. J. Appl. Mech. Tech. Phys. **53**, 517–531 (2012)
39. Grigor'ev, Yu.N., Ershov, I.V.: Linear stability of the Couette flow of a vibrationally excited gas. 1. Inviscid problem. J. Appl. Mech. Tech. Phys. **55**, 258–269 (2014)
40. Grigor'ev, Yu.N., Ershov, I.V.: Linear stability of the Couette flow of a vibrationally excited gas. 2. Viscous problem. J. Appl. Mech. Tech. Phys. **57**, 247–257 (2016)
41. Grigor'ev, Yu.N., Ershov, I.V.: Asymptotic theory of neutral stability curve of the Couette flow of a vibrationally excited gas. J. Phys.: Conf. Ser. **722**, 012012 (2016)
42. Grigor'ev, Yu.N., Ershov, I.V.: Linear stability of Couette flow of vibrationally non-equilibrium gas. AIP Conf. Proc. **1770**, 030043 (2016)

Chapter 1
Physico-Mathematical Models of Relaxing Molecular Gas Flows

Abstract The chapter has an introductory character and provides some auxiliary material to give an idea of notions and results of physical kinetics, kinetic theory, and acoustics of molecular gases, which are used in the book. The main goal of this chapter is to demonstrate the feasibility and adequacy of physical and mathematical models used in the authors' researches. In particular the evolution of the concept of bulk viscosity in mechanics and kinetic theory of gases is briefly described. Qualitative properties of the Landau-Teller relaxation equation for the vibrational mode energy, which plays a key role in subsequent considerations, are discussed. The physical mechanism of dissipation of acoustic waves on the background of the relaxation process in a thermally nonequilibrium molecular gas is explained.

The aim of the introductory chapter is to provide for specialists in aerodynamics some concepts and results of the kinetic theory of polyatomic gases, physical kinetics and acoustics, which are used in subsequent chapters. For this reason the material is briefly presented at a simple level without particular details. The chapter also includes some reference materials, which the readers may require for making their own quantitative estimates.

Section 1.1 contains basic information from the kinetic theory of thermal relaxation of molecular gases. A kinetic equation in the general form is provided, which takes into account elastic and inelastic collisions of gas molecules and describes the relaxing gas evolution at the level of the distribution function of molecules on their velocities and internal energy states. Expressions for gas-dynamic parameters in terms of the microscopic variables and distribution function are introduced.

Section 1.2 briefly describes the results of using the generalized Chapman–Enskog method for deriving gas-dynamic equations in the zeroth- and first-order approximations in order to demonstrate the degree of validity, the character of constraints, and the domains of applicability of these equations from the viewpoint of the kinetic theory of molecular gases, because these equations are used as a basis for further considerations. Some qualitative properties of the Landau–Teller relaxation equation are considered, which are important issues in subsequent chapters.

Section 1.3 presents a brief outline of the evolution of the notion of bulk viscosity in mechanics and in the kinetic theory of gases because this phenomenon still remains

Y.N. Grigoryev and I.V. Ershov, *Stability and Suppression of Turbulence in Relaxing Molecular Gas Flows*, Fluid Mechanics and Its Applications 117,
DOI 10.1007/978-3-319-55360-3_1

disputable in aerodynamics. Formulas and experimental data on the bulk viscosity coefficients and rotational and vibrational relaxation times for basic diatomic gases are given as reference materials. The authors' calculations of these parameters as functions of temperature are presented in graphical form, which provide a sufficiently grounded idea of the real ranges of these parameters for aerodynamic problems.

Section 1.4 is devoted to consideration of the physical mechanism of dissipation of acoustic oscillations on the background of the relaxation process in a thermally nonequilibrium molecular gas.

1.1 Elements of Physical Kinetics

The present consideration is confined to a single-component gas whose molecules have three translational, two rotational, and several vibrational degrees of freedom. In particular, molecules of diatomic gases, which are mainly considered in subsequent chapters, are characterized by one vibrational degree of freedom (vibrational mode). If the gas is thermally excited, collisional relaxation occurs, whose final result is the state of thermal equilibrium defined by the only temperature T. In this state the so-called equipartition of molecules in terms of their degrees of freedom arises: where the mean energy per one molecule in translational motion is $\varepsilon_t = 3\mathrm{k}T/2$, the mean energy of molecule rotation is $\varepsilon_r = \mathrm{k}T$, and the mean energy of one vibrational mode is $\varepsilon_v = \mathrm{k}T$, where k is the Boltzmann constant. Elastic and inelastic collisions are distinguished. In an elastic collision, which can be obviously illustrated by a collision of two smooth elastic spheres, the molecules exchange only kinetic energy and momentum. In inelastic collisions however the internal energy of molecules is changed. For example the energy of translational motion transforms to vibrational energy or energy transfer in the opposite direction occurs. The energy of translational motion of a molecule can take arbitrary values or, which is the same, translational motion has a continuous energy spectrum. Rotational and vibrational molecular energies can take only discrete quantum values, i.e., molecules have discrete energy spectra in their rotational and vibrational motions. A certain quantum number is assigned to each energy level. The mutual influence of rotational and translational states of molecules is usually assumed to be negligibly small.

Reaching the thermal equilibrium state in a gas depends on the intensity of energy exchange between individual degrees of freedom of molecules. Energy transitions between translational degrees of freedom of molecules without changes in the internal energy of rotational and vibrational modes (usually called TT-transitions) prevail in collisions [1–3]. They occur within times of the order of the mean free path time τ_{tt}, which serves as an estimate of the characteristic time of reaching equilibrium in terms of translational degrees of freedom. However, it is more difficult to reach the equilibrium state in terms of rotational degrees of freedom. It mainly occurs owing to energy exchange between translational and rotational modes (RT-exchange). A typical estimate of the time of this process is

$$\tau_{rt} = Z_r \tau_{tt}, \quad Z_r \simeq 10.$$

Energy exchange between translational and vibrational modes involves the greatest difficulties. At moderate temperatures of the flow, a transition to thermodynamic equilibrium requires several tens of thousands of collisions per one molecule. A typical time of VT-relaxation is

$$\tau_{vt} = Z_v \tau_{tt}, \quad Z_v \simeq 10^5\text{–}10^6.$$

In some cases, in particular in high-enthalpy flows, a transition to equilibrium in terms of vibrational modes occurs with an intermediate stage where the key role belongs to resonance transitions of vibrational energy where particle exchange vibrational quanta without changing the total vibrational energy (the so-called VV-transitions). The result of this process is the quasi-steady Boltzmann distribution over vibrational energy levels with a vibrational temperature T_v different from the static (translational) temperature of the flow T [4]. A typical time of this process is estimated as

$$\tau_{vv} = Z_{vv} \tau_{tt}, \quad Z_{vv} \simeq 10^3\text{–}10^4.$$

By comparing the above-mentioned characteristic times of relaxation processes with the characteristic time of the flow τ_f, one can define two types of hierarchy of relaxation times typical for aerodynamic problems. In the first case

$$\tau_{tt} \leq \tau_{rt} \ll \tau_{vt} \leq \tau_f. \tag{1.1}$$

The system of inequalities in the case of resonance VV-transitions is somewhat different:

$$\tau_{tt} \leq \tau_{rt} \ll \tau_{vv} \ll \tau_{vt} \leq \tau_f. \tag{1.2}$$

Owing to these relations, relaxing gas flows in these two cases can be described by different systems of hydrodynamic equations, which take into account the kinetics of relaxation processes with sufficient accuracy.

Derivation of systems of equations of relaxation hydrodynamics is based on kinetic equation of polyatomic gases, known as the Wang Chang–Uhlenbeck equation [4–6], which describe the space and time evolution of the distribution function of gas molecules over the spatial coordinates, translational velocities, and vibrational and rotational energy levels. Omitting detailed considerations of their structure, which is outside the scope of the present book, we can write the kinetic equation for a single-component gas in the absence of mass force fields as

$$\frac{\partial f_{ij}}{\partial t} + \mathbf{c} \cdot \frac{\partial f_{ij}}{\partial \mathbf{r}} = J_{ij}, \quad i = 0, \ldots, L_i, \quad j = 0, \ldots, L_{ij}. \tag{1.3}$$

Here f_{ij} is the distribution function of gas molecules, L_i is the number of excited vibrational levels, and L_{ij} is the number of excited rotational levels of molecules located at the ith vibrational level. The right-hand side of this equation contains the integral operator J_{ij}, which describes the change in the distribution function with fixed values of all variables in collisions of different types. It has the following components corresponding to the above-introduced exchange processes:

$$J_{ij} = J_{ij}^{TT} + J_{ij}^{RT} + J_{ij}^{VV} + J_{ij}^{VT}. \tag{1.4}$$

The distribution function $f_{ij}(\mathbf{c}, \mathbf{r}, \varepsilon_{rij}, \varepsilon_{vi}, t)$ has the meaning of the number density of molecules at the time t in the volume element $d\mathbf{r}$ with the spatial coordinate $\mathbf{r}$ and velocity in the interval $d\mathbf{c}$ around $\mathbf{c}$, which are located at the vibrational level i and rotational level j. All macroscopic characteristics of the gas are expressed via the distribution function, and the relationship between the kinetic and gasdynamic descriptions is established. They are calculated as power moments of the distribution function by means of integration over the entire velocity space. Summation is performed over all energy states of the molecules. In particular the density of the gas with a molecular mass m is expressed as

$$\rho(\mathbf{r}, t) = nm = m \sum_{ij} \int f_{ij}\, d\mathbf{c}, \tag{1.5}$$

where n is the number of gas molecules in a unit volume.

The macroscopic momentum of a unit volume of the gas is

$$\rho\,\mathbf{u} = m \sum_{ij} \int \mathbf{c} f_{ij}\, d\mathbf{c}, \tag{1.6}$$

where $\mathbf{u}(\mathbf{r}, t)$ is the hydrodynamic velocity of the flow.

The total density of the internal energy of the gas is defined by the expression

$$\rho\, U(\mathbf{r}, t) = \rho\,(E_t + E_r + E_v)$$

$$= \sum_{ij} \int \frac{m\,\mathbf{C}^2}{2} f_{ij}\, d\mathbf{c} + \sum_{ij} \varepsilon_{rij} \int f_{ij}\, d\mathbf{c} + \sum_{ij} \varepsilon_{vi} \int f_{ij}\, d\mathbf{c}. \tag{1.7}$$

Here E_t, E_r, and E_v are the translational, rotational, and vibrational energies of a unit mass of the gas, ε_{rij} is the rotational energy in the ijth state, ε_{vi} is the vibrational energy of the ith vibrational level, and $\mathbf{C} = \mathbf{c} - \mathbf{u}$ is the thermal velocity of molecules.

The stress tensor $P(\mathbf{r}, \mathrm{t})$ is calculated as

$$P = m \sum_{ij} \int \mathbf{C}\,\mathbf{C} f_{ij}\, d\mathbf{c}, \tag{1.8}$$

where $\mathbf{C}\,\mathbf{C}$ is the tensor of the thermal velocities of molecules.

The total heat flux $\mathbf{q}(\mathbf{r}, t)$ consists of the translational, rotational, and vibrational energy fluxes and is written as

$$\mathbf{q} = \mathbf{q}_t + \mathbf{q}_r + \mathbf{q}_v$$

$$= \sum_{ij} \int \mathbf{C} \frac{m\mathbf{C}^2}{2} f_{ij}\, d\mathbf{c} + \sum_{ij} \varepsilon_{rij} \int \mathbf{C} f_{ij}\, d\mathbf{c} + \sum_{ij} \int \varepsilon_{vi} \mathbf{C} f_{ij}\, d\mathbf{c}. \tag{1.9}$$

1.2 Systems of Equations of Relaxation Gas Dynamics

The basic goal of this section is to provide the minimum idea about the degree of theoretical justification of equations used in subsequent chapters, as well as the limits of their applicability, because this area of aeromechanics is still far from being a desktop tutorial.

The above-introduced relations (1.5)–(1.9) between the kinetic characteristics of a nonequilibrium molecular gas and gas-dynamic variables allow us to illustrate how the kinetic theory can be applied to derive gas-dynamic equations of a relaxing gas for different levels of excitation of internal degrees of freedom.

Special normalization converts the kinetic equation (1.3) to

$$\frac{\partial \bar{f}_{ij}}{\partial \bar{t}} + \mathbf{c} \cdot \frac{\partial \bar{f}_{ij}}{\partial \bar{\mathbf{r}}} = \bar{J}_{ij} = \frac{1}{\delta_1} \bar{J}_{ij}^{TT} + \frac{1}{\delta_2} \bar{J}_{ij}^{RT} + \frac{1}{\delta_3} \bar{J}_{ij}^{VV} + \frac{1}{\delta_4} \bar{J}_{ij}^{VT}, \tag{1.10}$$

$$i = 0, \ldots, L_i, \quad j = 0, \ldots, L_{ij},$$

where

$$\delta_1 = \tau_{tt}/\tau_f, \quad \delta_2 = \tau_{rt}/\tau_f, \quad \delta_3 = \tau_{vv}/\tau_f, \quad \delta_1 = \tau_{vt}/\tau_f. \tag{1.11}$$

The sum of the molecular masses, the total momentum, and the total energy of colliding particles obviously remain unchanged in each collision of a pair of molecules. In the case of resonance VV-exchange the sum of vibrational quanta is also unchanged. Therefore the quantities

$$m, \quad m\mathbf{c}, \quad \frac{m\mathbf{C}^2}{2} + \varepsilon_{rij} + \varepsilon_{vi},$$

and also ε_{vi} are called the collision (summator) invariants of molecular collisions [5].

Using the kinetic equation (1.10), multiplying it by collision invariants, integrating over the velocity space, and summing over all energy states in accordance with definitions (1.5)–(1.9), we derive the general system of mass, momentum, and total energy conservation equations:

$$\frac{\partial \rho}{\partial t} + \frac{\partial \rho u_i}{\partial x_i} = 0, \quad \rho\left(\frac{\partial u_i}{\partial t} + u_j \frac{\partial u_i}{\partial x_j}\right) + \frac{\partial P_{ij}}{\partial x_i} = 0,$$

$$\rho\left(\frac{\partial U}{\partial t} + u_j \frac{\partial U}{\partial x_j}\right) + \frac{\partial q_i}{\partial x_i} + P_{ij} \frac{\partial u_i}{\partial x_j} = 0. \tag{1.12}$$

If there is resonance VV-exchange (see Eq. (1.2)), system (1.12) is supplemented with the equation for vibrational energy, which has the following form [4]:

$$\rho\left(\frac{\partial E_v}{\partial t} + u_j \frac{\partial E_v}{\partial x_j}\right) + \frac{\partial q_{vi}}{\partial x_i} = \sum_{ij} \varepsilon_{vi} \int \bar{J}_{ij}^{VT} d\mathbf{c}. \tag{1.13}$$

Here the subscripts of the spatial and hydrodynamic variables denote the projections of the vector variables onto the corresponding axes of the Cartesian coordinate system, and summation is performed over repeated subscripts. The zero values of the right-hand sides of the mass, momentum, and total energy conservation equations are obtained by integration of the integral operators in Eq. (1.10) with corresponding weights from the collision invariants. This is related to the fundamental properties of the symmetry of Boltzmann-type collision integrals [5].

Using closures of different levels of system (1.12), (1.13) one can derive different systems of relaxation gas dynamics. The closure is understood as obtaining expressions for the stress tensor $\boldsymbol{P}$ and the heat flux vector $\mathbf{q}$ in terms of the gradients of hydrodynamic and thermodynamic variables, as well as detalization of the right-hand side of the equation for vibrational energy (1.13). The generalized Chapman–Enskog method was developed for this purpose in the kinetic theory of polyatomic gases [1] (see also the references in [4, 5]).

The hierarchy of the kinetic times (1.1), (1.2) corresponds to the hierarchy of the small parameters (1.11). It is seen that they are included into the right-hand sides of system (1.10) in a singular manner. This allows the solutions of Eq. (1.10) to be found using an asymptotic method of the perturbation theory, by which the generalized Chapman–Enskog method is presented from the mathematical viewpoint. Thus the systems of equations of relaxation gas dynamics obtained by this method, which are considered below, are usually considered as asymptotic solutions of Eq. (1.10) [1].

1.2.1 One-Temperature Models of the Flow

Vibrational degrees of freedom are basically non-excited at small deviations from thermodynamic equilibrium. In this case the internal energy of molecules is concentrated mostly in translational and rotational degrees of freedom. Moreover rotational modes become excited already at temperatures higher than absolute zero only by several degrees. Therefore, both rotational and translational motions can be assumed to be always present in molecular gas flows of interest for us.

In this case summation over vibrational levels and contributions of vibrational modes to the total energy U and heat flux $\mathbf{q}$ should be omitted in Eqs. (1.5)–(1.9). The equation for vibrational energy (1.13) vanishes in system (1.12), and the remaining equations retain their form.

The integral operators $\bar{J}_{ij}^{VV}$ and $\bar{J}_{ij}^{VT}$ are eliminated from the kinetic equation (1.10), as well as the dependence of the distribution function on the vibrational level of number i. By virtue of the left inequality in relations (1.1), (1.2), it may be assumed that

$$\delta_1 \sim \delta_2 \simeq \varepsilon.$$

The thus-adapted kinetic equation has only one small parameter, and it takes the form

$$\frac{\partial \bar{f}_j}{\partial \bar{t}} + \mathbf{c} \cdot \frac{\partial \bar{f}_j}{\partial \bar{\mathbf{r}}} = \frac{1}{\varepsilon} \bar{J}_j = \frac{1}{\varepsilon} \left(\bar{J}_{ij}^{TT} + \bar{J}_{ij}^{RT} \right). \tag{1.14}$$

The solution of Eq. (1.14) is sought in the form of the Chapman–Enskog series

$$\bar{f}_j(\bar{\mathbf{c}}, \bar{\mathbf{r}}, \varepsilon_{rj}, \bar{t}) = \sum_k \varepsilon^k \bar{f}_j^k (\bar{\mathbf{c}}, U, \nabla \bar{\mathbf{c}}, \ldots).$$

A specific feature of the Chapman–Enskog expansion is the fact that the dependence of the functions $\bar{f}_j^k$ of all approximations on time and spatial coordinates is realized exclusively via the gas macroparameters and their spatial gradients of all orders.

Substitution yields the integral equation for the zero-order approximation function

$$\bar{J}_{j,0}^{TT} + \bar{J}_{j,0}^{RT} = 0.$$

Its solution has the form

$$\bar{f}_j^0 = n \left(\frac{m}{2\pi \,\mathrm{k} T} \right)^{3/2} \exp\left(-\frac{m\mathbf{C}^2}{2\mathrm{k} T} \right) N_r \exp\left(-\frac{\varepsilon_j}{\mathrm{k} T} \right), \tag{1.15}$$

where k is the Boltzmann constant and N_r is the normalization factor of the rotational energy distribution. Thus, the distribution function in the zero-order approximation is the product of the Maxwell distribution over translational velocities and the Boltzmann distribution over rotational energy levels with a single temperature T.

In accordance with the condition adopted in the Chapman–Enskog method [1] (see also [4, 5]) the gas macroparameters (1.5)–(1.7) are completely determined by the distribution functions of the zero-order approximation $\bar{f}_j^0$. Subsequent approximations contribute only to the stress tensor (1.8) and heat flux (1.9). As f_j^0 is an even function, substitution of Eq. (1.15) into these relations yields

$$P = n\mathrm{k}T\,\mathbf{I}, \quad \mathbf{q} = 0,$$

where **I** is the unit tensor of the second rank.

As a result the macroscopic equations of conservation (1.12) transform to a usual system of gas-dynamic equations for an inviscid nonheat-conducting gas [4, 7, 8]:

$$\frac{\partial \rho}{\partial t} + \frac{\partial \rho u_i}{\partial x_i} = 0,$$

$$\rho \left(\frac{\partial u_i}{\partial t} + u_j \frac{\partial u_i}{\partial x_j} \right) = -\frac{\partial p}{\partial x_i},$$

$$\rho \left(\frac{\partial U_1}{\partial t} + u_j \frac{\partial U_1}{\partial x_j} \right) + p \frac{\partial u_i}{\partial x_i} = 0, \quad p = n\mathrm{k}T, \tag{1.16}$$

where p is the hydrostatic pressure, and the internal energy of the gas is determined as

$$U_1 = \frac{3\mathrm{k}}{2m} T + E_r/\rho = \frac{5}{2} \frac{\mathrm{k}\, T}{m}.$$

In view of the form of the zero-order approximation function (1.15) and equations (1.16) for the function in the first-order approximation $\bar{f}_j^1$, the kinetic equation (1.14) yields a linear integral equation. Its solution has the following structural form [4, 5]:

$$\bar{f}_j^1 = f_j^0 \frac{1}{n} \left(-\mathbf{A}_j \cdot \frac{\partial \ln T}{\partial \mathbf{r}} - \mathbf{B}_j : \frac{\partial \mathbf{u}}{\partial \mathbf{r}} - G_j \operatorname{div} \mathbf{u} \right). \tag{1.17}$$

The vector function $\mathbf{A}_j$, the tensor function $\mathbf{B}_j$, and the scalar function G_j in Eq. (1.17), which depend only on the thermal velocities of molecules, are calculated explicitly. They are used to express the heat flux and corrections to the stress tensor. The corresponding expressions are written as

$$q_{is} = -(\lambda_t + \lambda_r) \frac{\partial T}{\partial x_i}, \tag{1.18}$$

$$P_{ij} = p\, \delta_{ij} - \eta \left(\frac{\partial u_i}{\partial x_j} + \frac{\partial u_j}{\partial x_i} - \frac{2}{3} \frac{\partial u_i}{\partial x_i} \right) - \eta_b \frac{\partial u_k}{\partial x_k} \delta_{ij} = p\, \delta_{ij} - p_{ij}. \tag{1.19}$$

Here λ_t and λ_r are the thermal conductivities due to translational and rotational modes of gas molecules, respectively, η and η_b are the shear and bulk viscosity coefficients, and p_{ij} is the viscous stress tensor.

It is shown in the kinetic theory of gases that the bulk viscosity coefficient is calculated by the formula [5]

$$\eta_b = \left(\frac{\mathrm{k}\, c_V^{int}}{m\, c_V^2} \right) p\, \tau_{int}, \tag{1.20}$$

where c_V^{int} is the heat capacity at a constant volume of internal degrees of freedom of molecules, c_V is the total heat capacity at a constant volume, and τ_{int} is the relaxation time of the corresponding inelastic process.

In this case $c_V^{int} = c_{V,r}$ is related to the energy of rotational degrees of freedom, $c_V = 3\text{k}/(2m) + c_{V,r}$, and the relaxation time $\tau_{int} = \tau_{rt}$ is the characteristic time of RT-exchange.

Thus, the emergence of the contribution of bulk viscosity in the stress tensor and its relationship with the relaxation process is a rigorous result of the kinetic theory of polyatomic gases. It should be noted that bulk viscosity in monatomic (e.g., inert) gases is equal to zero because they have no internal degrees of freedom and $c_V^{int} = 0$. The effect of bulk viscosity was explained for the first time within the framework of the kinetic theory by Leontovich [9, 10], though the theory of polyatomic gases was not yet completely established at that time.

Substitution of Eqs. (1.18), (1.19) into the global system of macroscopic equations of conservation (1.12) yields a system of Navier–Stokes equations for a viscous heat-conducting gas where the relaxation process is taken into account by the term with bulk viscosity in the stress tensor:

$$\frac{\partial \rho}{\partial t} + \frac{\partial \rho u_i}{\partial x_i} = 0, \quad \rho\left(\frac{\partial u_i}{\partial t} + u_j \frac{\partial u_i}{\partial x_j}\right) = -\frac{\partial p}{\partial x_i} - \frac{\partial p_{ij}}{\partial x_j},$$

$$\rho\left(\frac{\partial U_1}{\partial t} + u_j \frac{\partial U_1}{\partial x_j}\right) + p\frac{\partial u_i}{\partial x_i} = -\frac{\partial q_{is}}{\partial x_i} - p_{ij}\frac{\partial u_i}{\partial x_j}, \quad p = n\text{k}T. \tag{1.21}$$

1.2.2 Two-Temperature Models of Relaxing Flows

It was shown in [1–4, 11, 12] that collisions of particles in many vibrationally excited high-enthalpy gas flows lead to much more frequent events of vibrational energy quanta exchange than events of vibrational energy conversion to translational and rotational energy. Therefore, the key role in relaxation to the equilibrium state in this case belongs to inelastic VV-exchange events, which occur at characteristic time τ_{vv}. Moreover, the times of relaxation processes satisfy the system of inequalities (1.2). Such flows occur in the working tract of gas dynamic lasers, in supersonic nozzles and underexpanded jets, behind the fronts of strong shock waves, and in thermally nonequilibrium boundary layers.

The difference in the characteristic times of more rapid processes for this class of flows at the time scales

$$\tau_{vt} \leq \tau_f$$

can be considered to be negligibly small. This fact allows one to assume that

$$\delta_1 \sim \delta_2 \sim \delta_3 \sim \varepsilon_1, \quad \delta_4 \sim 1$$

in the kinetic equation (1.10). Therefore, this equation can be rewritten as

$$\frac{\partial \bar{f}_{ij}}{\partial \bar{t}} + \mathbf{c}\cdot\frac{\partial \bar{f}_{ij}}{\partial \bar{\mathbf{r}}} = \bar{J}_{ij} = \frac{1}{\varepsilon_1}\left(\bar{J}_{ij}^{TT} + \bar{J}_{ij}^{RT} + \bar{J}_{ij}^{VV}\right) + \bar{J}_{ij}^{VT} \tag{1.22}$$

with the only singular small parameter ε_1. The solution of Eq. (1.22) is sought in the form of the Chapman–Enskog series, which can be written as [4]

$$\bar{f}_{ij}(\bar{\mathbf{c}},\ \bar{\mathbf{r}},\ \varepsilon_{rij},\ \varepsilon_{vi},\ \bar{t}) = \sum_k \varepsilon_1^k \bar{f}_{ij}^k\,(\bar{\mathbf{c}},\ U,\ \nabla\bar{\mathbf{c}},\ \ldots)\,.$$

Substitution of this relation into Eq. (1.22) yields the integral equation for the distribution function in the zero-order approximation:

$$\bar{J}_{ij,0}^{TT} + \bar{J}_{ij,0}^{RT} + \bar{J}_{ij,0}^{VV} = 0.$$

Its unique solution has the form

$$\bar{f}_{ij}^0 = n\left(\frac{m}{2\pi\,\mathrm{k}T}\right)^{3/2} \exp\left(-\frac{m\mathbf{C}^2}{2\mathrm{k}T}\right) N_r \exp\left(-\frac{\varepsilon_j}{\mathrm{k}T}\right) N_v \exp\left(-\frac{\varepsilon_i}{\mathrm{k}T_v}\right), \tag{1.23}$$

where N_v is the normalization factor of the vibrational energy distribution. In contrast to Eq. (1.15) the distribution function of the zero-order approximation includes the quasi-equilibrium Boltzmann distribution over vibrational energy levels with a vibrational temperature T_v in addition to the product of the Maxwell distribution over translational velocities and the Boltzmann distribution over rotational energy levels with a unified temperature T.

Similar to distribution (1.15), $\bar{f}_{ij}^0$ is an even function of thermal velocities. Therefore, its substitution into the expressions for the stress tensor (1.8) and heat flux tensor (1.9) yields

$$P = n\mathrm{k}T\,\mathbf{I}, \qquad \mathbf{q} = 0,$$

$$U = \frac{3\mathrm{k}}{2m}T + E_r/\rho + E_v/\rho = \frac{5}{2}\frac{\mathrm{k}T}{m} + 1/\rho\sum_i \varepsilon_i n_i. \tag{1.24}$$

In addition to conditions (1.5) and (1.7), the following normalization condition is used for the populations of vibrational levels n_i in the generalized Chapman–Enskog method [4]:

$$n_i = \sum_j \int f_{ij}^0\, d\mathbf{c}. \tag{1.25}$$

For obtaining a system of equations for the dynamics of an inviscid non-heat-conducting vibrationally excited gas from Eq. (1.12) by using the zero-order approximation (1.23), it is necessary to calculate the right-hand side of Eq. (1.13) for vibrational energy E_v, which in turn requires the knowledge of the populations n_i of excited vibrational levels. To simplify calculations it is usually assumed [4–6], that molecular vibrations are described by the harmonic oscillator model. In this case the vibrational energy of a molecule at the ith energy level is $\varepsilon_{vi} = i\Delta\varepsilon$, where $\Delta\varepsilon$ is the vibrational quantum. Moreover, all vibrational levels are equidistant, the number of vibrational levels is infinite, and only single-quantum transitions between the nearest levels are allowed. Calculations performed under these assumptions yield the Landau–Teller equation

$$\frac{\partial E_v(T_v)}{\partial t} + u_i \cdot \frac{\partial E_v(T_v)}{\partial x_i} = -\frac{E_v(T_v) - E_v^{eq}(T)}{\tau_{vt}(n, T)}, \tag{1.26}$$

where the equilibrium vibrational energy of molecules is

$$E_v^{eq}(T) = E_v(T) = N\mathrm{k}T/m,$$

N is the number of vibrational degrees of freedom of the molecule.

The harmonic oscillator model describes the real vibrational kinetics only if a comparatively small number of lower vibrational levels are excited. In practice however, the area of applicability of the Landau–Teller equation is extended with using experimental data on relaxation times [2, 8]. Deep excitation with many levels involved leads to anharmonicity: the upper energy levels are no longer equidistant and become more condensed, whereas the energy exchange between the levels becomes inhomogeneous along the spectrum. Nevertheless, it is possible to use an empirical correction to the relaxation time [2, 12], which takes into account anharmonic effects, if consideration is confined to temperatures that ensure insignificant dissociation of molecules.

Substituting the relations of the zero-order approximation (1.24) into system (1.12) and subtracting the Landau–Teller equation from the energy equation, we obtain a system of equations of the dynamics of a vibrationally excited inviscid nonheat-conducting gas in the following form:

$$\frac{\partial \rho}{\partial t} + \frac{\partial \rho u_i}{\partial x_i} = 0, \quad \rho\left(\frac{\partial u_i}{\partial t} + u_j \frac{\partial u_i}{\partial x_j}\right) = -\frac{\partial p}{\partial x_i},$$

$$\rho\left(\frac{\partial U_1}{\partial t} + u_j \frac{\partial U_1}{\partial x_j}\right) + p\frac{\partial u_i}{\partial x_i} = \rho\frac{E_v(T_v) - E_v^{eq}(T)}{\tau_{vt}(n, T)},$$

$$\frac{\partial E_v(T_v)}{\partial t} + u_i \cdot \frac{\partial E_v(T_v)}{\partial x_i} = -\frac{E_v(T_v) - E_v^{eq}(T)}{\tau_{vt}(n, T)}, \quad p = n\mathrm{k}T. \tag{1.27}$$

Using the zero-order approximation function (1.23) and eliminating the derivatives with respect to time with the use of system (1.27), it is possible to obtain a nonhomogeneous integral equation for the distribution function of the first-order approximation from Eq. (1.22). Similar to (1.17) its solution can be presented as

$$\bar{f}_{ij}^1 = f_{ij}^0 \frac{1}{n}\left(-\mathbf{A}_{ij} \cdot \frac{\partial \ln T}{\partial \mathbf{r}} - \mathbf{A}_{ij}^v \cdot \frac{\partial \ln T_v}{\partial \mathbf{r}} - \mathbf{B}_j : \frac{\partial \mathbf{u}}{\partial \mathbf{r}} - G_j \operatorname{div} \mathbf{u}\right). \tag{1.28}$$

Here an additional term appears which describes energy transfer by excited vibrational degrees of freedom.

When solution (1.28) is found explicitly, the correction to the stress tensor is calculated:

$$P_{ij} = p\,\delta_{ij} - \eta\left(\frac{\partial u_i}{\partial x_j} + \frac{\partial u_j}{\partial x_i} - \frac{2}{3}\frac{\partial u_k}{\partial x_k}\right) - \eta_b\,\frac{\partial u_i}{\partial x_i}\delta_{ij} = p\,\delta_{ij} - p_{ij}. \tag{1.29}$$

In contrast to Eq. (1.19), the bulk viscosity coefficient is presented here as a sum of two terms [4]:

$$\eta_b = \eta_{b\,r} + \eta_{b\,v}.$$

The first term is related to inelastic TR-exchange, and the second term is associated with resonance VV-exchange.

Correspondingly, the expression for the heat flux is

$$q_i = q_{is} + q_{iv}, \quad q_{is} = -(\lambda_t + \lambda_r)\,\frac{\partial T}{\partial x_i} \qquad q_{iv} = -\lambda_v\frac{\partial T_v}{\partial x_i}, \tag{1.30}$$

where λ_v is the thermal conductivity associated with vibrational motion, and the flux q_{iv} describes diffusion transport of energy contained in vibrational modes.

It should be noted that the function of the first approximation (1.28) also yields a correction to the relaxation term in the right-hand side of the equation for E_v in system (1.12). As was demonstrated in [13], if this correction is taken into account, it is impossible to pass rigorously to the Landau–Teller equation even in the model of harmonic oscillations. Nevertheless, it can be concluded from the results calculated in [13] that this correction is essential at $T \simeq T_v \approx 5000$ K. This range of static temperatures T is not of interest for flows considered in this book. Therefore, this correction is ignored below.

In view of Eqs. (1.29) and (1.30) and the remark made above, the system of equations of two-temperature aerodynamics of a viscous heat-conducting vibrationally excited gas is presented as

$$\frac{\partial \rho}{\partial t} + \frac{\partial \rho u_i}{\partial x_i} = 0, \quad \rho\left(\frac{\partial u_i}{\partial t} + u_j\,\frac{\partial u_i}{\partial x_j}\right) = -\frac{\partial p}{\partial x_i} + \frac{\partial p_{ij}}{\partial x_j},$$

$$\rho\left(\frac{\partial U_1}{\partial t} + u_j\,\frac{\partial U_1}{\partial x_j}\right) + p\,\frac{\partial u_i}{\partial x_i} = -\frac{\partial q_{is}}{\partial x_i} - p_{ij}\frac{\partial u_i}{\partial x_j} + \rho\frac{E_v(T_v) - E_v^{eq}(T)}{\tau_{vt}(n,\,T)},$$

$$\rho\left(\frac{\partial\,E_v(T_v)}{\partial t} + u_i\cdot\frac{\partial\,E_v(T_v)}{\partial x_i}\right) = -\frac{\partial q_{iv}}{\partial x_i} - \rho\frac{E_v(T_v) - E_v^{eq}(T)}{\tau_{vt}(n,\,T)},$$

$$p = n\mathrm{k}T. \tag{1.31}$$

1.2.3 Landau–Teller Equation

The Landau–Teller equation occupies a special place among mathematical models used in subsequent considerations. Moreover, this equation was little used until recently not only in the hydrodynamic stability theory, but also in aerodynamic problems in general. For this reason it seems necessary to discuss the character of the relaxation process described by this equation and its mathematical properties.

Omitting the subscripts, which are not important here, we consider Eq. (1.26) in the form

$$\frac{dE_v(t, T_v)}{dt} = \frac{E_v(T) - E_v(t, T_v)}{\tau}. \tag{1.32}$$

This form provides a description of the relaxation process at each point of the flow with the transport process being ignored or corresponds to extremely fast relaxation where this process can be neglected. The properties that deserve attention remain unchanged. Being a typical relaxation equation, Eq. (1.32) shows that its solution always tends to the equilibrium state characterized by a unified temperature T. The instantaneous velocity of tending to the equilibrium state is directly proportional to the deviation from the equilibrium solution at each time instant. At $T \simeq$ const the solution of Eq. (1.32) has the form

$$E_v(t, T_v) - E_v(T) = \left[E_v(0, T_v) - E_v(T)\right] \exp\left(-t/\tau\right).$$

It follows from here that convergence to the equilibrium solution has an exponential character.

The process of vibrational relaxation is irreversible and is accompanied by an increase in entropy. This property is reflected in Eq. (1.32). Indeed, as was shown in [14], the increase in entropy in the process of vibrational relaxation with allowance for Eq. (1.32) is expressed by the formula

$$\Delta S = \left(\frac{1}{T_v} - \frac{1}{T}\right) \Delta E_v$$

It is seen that the change in entropy is equal to zero only in two extreme cases: either the equilibrium state where $T = T_v$ or an intermediate "frozen" state in an adiabatic process where $T \neq T_v$, but $\Delta E_v = 0$. In the general case it follows from the formula that the increase in entropy is always nonnegative ($\Delta S \geq 0$) for the difference of vibrational energy with an arbitrary sign ($\Delta E_v < 0$ or $\Delta E_v > 0$). Thus, if $\Delta E_v > 0$, then the difference of the inverse temperatures in brackets is also positive. Vice versa, if $\Delta E_v < 0$, then $T < T_v$ and the expression in brackets is negative. It follows from here in particular that the dissipative effect of vibrational relaxation occurs both in an underexpanded supersonic jet or laser pumping of the vibrational mode where $T < T_v$ and behind shock waves where $T > T_v$.

It is useful to note that it is possible to avoid including the Landau–Teller equation for each vibrational mode in the first approximation in considering the flow of a polyatomic gas (e.g., carbon dioxide). Instead, it is possible to use a single-mode description where the "effective" relaxation time τ_e is defined as

$$\frac{1}{\tau_e} = \frac{c_v}{c_1}\frac{1}{\tau_1},$$

where c_v is the specific heat of the gas at constant volume, and c_1 and τ_1 are the heat capacity and relaxation time of the most rapidly relaxing mode. In the flow of a gas mixture it is possible to pass to the single-mode description by using the effective relaxation time "weighted" over the volume fractions of the components in accordance with the formula

$$\frac{1}{\tau_e} = \sum_i \frac{\xi_i}{\tau_i}, \quad \Sigma_i \xi_i = 1.$$

As was noted above, the relaxation equation was derived in original work [15] by semiempirical manner under assumptions that the vibrational mode is harmonic and the deviation from equilibrium is small at $T \simeq T_v$. The following relaxation equation was obtained in a recent publication [16] on the basis of the kinetic theory [4] under the same assumptions:

$$\frac{dE_v(t, T_v)}{dt} = \frac{T}{T_v}\frac{\left[E_v(T) - E_v(t, T_v)\right]}{\tau}. \tag{1.33}$$

As compared to rigorous kinetic calculations, it was shown [16] that Eq. (1.32) understates the vibrational relaxation rate $R_v = dE_v(t, T_v)/dt$ for diatomic gases at $T > T_v$ (e.g., behind shock waves). In the opposite case of expanding flows at $T < T_v$ Eq. (1.32) overstates the value of R_v. At the same time Eq. (1.33) provides an adequate approximation to kinetic calculations and authors of [16] recommend using this equation as a simple correction to the Landau–Teller equation in the case of a significant deviation from equilibrium, but for harmonic oscillations. At the same time it is seen from the same calculations that a more important factor that should be taken into account for chosen static temperature of the flow and temperature of the first vibrational level is the anharmonicity, which is not taken into account by the above-introduced correction. Therefore, to decide whether Eq. (1.33) is useful, it should be tested in real applications.

It is more important to use a correction of the Landau–Teller equation for taking into account the anharmonic character of molecule vibrations, which can be manifested, e.g., in underexpanded supersonic jets. This can be done by applying a phenomenological correction to the VT-relaxation time [2]. The correction is expressed by the formula

$$\tau_{vt}^{anh}(T, T_v) = \tau_{vt}(T)\left[\frac{1 - \gamma \exp(-\Theta/T_v)}{1 - \exp(-\Theta/T_v)}\right]^2. \tag{1.34}$$

Here Θ [K] is the characteristic temperature (energy of the basic vibrational quantum in Kelvin degrees). The thus-corrected time of VT-relaxation depends not only on the gas flow temperature T, but also on the vibrational temperature T_v.

It is of interest to estimate the upper limit of vibrational temperatures T_v where the correction for the anharmonicity of oscillations can be used. An obvious estimate of the limit of applicability of formula (1.34) in terms of T_v follows from the condition

$$1 - \gamma \exp(-\Theta/T_v) > 0,$$

whence

$$T_v < \Theta / \ln \gamma.$$

For N_2, O_2, and CO molecules, we have the approximate expression [2]

$$\gamma \approx 1.87\alpha\, \Theta^{2/3} T^{-1/3},$$

where $\alpha = \Theta/(4 E_{diss})$ is the anharmonicity parameter, E_{diss} [K] is the energy of dissociation of the molecule. Using the mean values of these parameters $\Theta \approx 2500$ K and $E_{diss} \approx 10^5$ K, we see that the correction for the anharmonicity of molecular vibrations in the case with $T \approx 2000$ K is applicable in the interval $T_v < 1.5 \times 10^4$ K.

For the same mean values of the parameters Θ, E_{diss}, and $T_v \approx \Theta$, the approximate value of the correction for the anharmonicity is

$$\left[\frac{1 - \gamma \exp(-\Theta/T_v)}{1 - \exp(-\Theta/T_v)}\right]^2 \approx \left[\frac{1 - 1.18\, e^{-1}}{1 - e^{-1}}\right]^2 \approx 0.82.$$

It may be assumed that this correction is within the range of scattering of experimental data in Fig. 1.7.

1.3 Parameters of Thermal Relaxation in Diatomic Gases

Among the characteristics of thermal relaxation process in molecular gas flows, the phenomenon of bulk viscosity is still a disputable issue. Introduction of the vibrational energy E_v as an additional macroparameter in the description of vibrationally excited gas flows does not inspire any significant objections. Nevertheless, there is no unique opinion about the use of the bulk (second) viscosity coefficient, up to completely unacceptable, both in manuals on hydrodynamics (see, e.g., [17]) and in respected periodic academic journals [18]. The authors' publications collected in this book demonstrate that bulk viscosity may produce a significant effect on stability of relaxing gas flows. Here we do not mean dissipation of far ultrasound, which is confirmed by numerous experiments (see [19] and the list of recent publications in [4]). The effect is manifested as suppression of hydrodynamic disturbances whose characteristic frequencies is lower at least by one order.

In this connection, it is desirable to provide a brief outline of the evolution of the bulk viscosity concept up to the modern idea about this phenomenon. The aim of this description, which is also the aim of the entire book to some extent, is to prevent of specialists in hydrodynamics from the negative attitude to bulk viscosity and arbitrary use the Stokes relation [17], which allows one to eliminate bulk viscosity from consideration.

1.3.1 Bulk Viscosity Phenomenon

The emergence of an additional viscosity coefficient in the classical hydrodynamics is associated with the Newton-Cauchy-Poisson phenomenological formula [17, 20], where the stress tensor is expressed via the strain rate tensor as

$$P_{ij}^{f} = p\,\delta_{ij} - \eta\left(\frac{\partial u_i}{\partial x_j} + \frac{\partial u_j}{\partial x_i}\right) - \lambda\,\frac{\partial u_k}{\partial x_k}\delta_{ij}. \tag{1.35}$$

Here η and λ are the originally empirical constants; η is associated with the shear viscosity coefficient, and λ is called as the second viscosity coefficient.

It is known that Eq. (1.35) is based on two postulates:

I. If there are no viscous effects, the stress tensor should coincide with the stress tensor in an ideal fluid (gas).

II. Additional terms (in addition to hydrostatic pressure) of the stress tensor are linear homogeneous functions of the strain rate tensor components, and the coefficients η and λ of these functions are independent of choosing the coordinate system.

In the kinetic theory of polyatomic gases [4–6] the stress tensor is usually written with the use of the non-divergent strain rate tensor (1.19), (1.29):

$$P_{ij} = p\,\delta_{ij} - \eta\left(\frac{\partial u_i}{\partial x_j} + \frac{\partial u_j}{\partial x_i} - \frac{2}{3}\frac{\partial u_i}{\partial x_i}\right) - \eta_b\,\frac{\partial u_k}{\partial x_k}\delta_{ij}. \tag{1.36}$$

Obviously, Eqs. (1.19), (1.29) and (1.36) derived within the framework of the generalized Chapman–Enskog method also satisfy the postulates of the phenomenological theory. In contrast to the latter, however, the kinetic theory allows derivation of expressions for calculating the coefficients η and η_b [4, 5]. This is one of the main achievements of the kinetic theory and serves as a confirmation of its adequacy. Comparing Eqs. (1.35) and (1.36), we find the relationship between the coefficients of bulk and second viscosities in the form

$$\eta_b = \frac{1}{3}\left(2\eta + 3\lambda\right). \tag{1.37}$$

The divergence (trace) of the phenomenological stress tensor (1.35) is expressed as

$$3p^m \equiv (P_{11} + P_{22} + P_{33}) = 3p - (2\eta + 3\lambda)\frac{\partial u_k}{\partial x_k}, \tag{1.38}$$

where the mean pressure p^m is introduced. This relation has served as a starting point in discussions about the second viscosity for almost a century [18, 19, 21].

In classical hydrodynamics [17, 20] it is usually proposed to avoid using the second viscosity coefficient in the following manner. In an incompressible fluid the velocity divergence is $\partial u_k/\partial x_k = 0$ and the mean pressure is equal to the hydrostatic pressure similar to the case of an ideal fluid (gas): $p^m = p$. For this relation to be valid in the general case, including compressible gases, it is necessary to set

$$2\eta + 3\lambda = 0$$

in Eq. (1.38), which is the known Stokes relation. It follows from Eq. (1.37) that it is equivalent to a zero value of bulk viscosity.

We cannot argue that this postulate remains "set in stone" for specialists in aerodynamics up to now, though it is still used as a convenient model in various aspects even in those cases where the necessity of taking into account the real gas effects is obvious.

Nevertheless, leading specialists in mechanics realized by the mid-1950s after numerous experiments on absorption and propagation of ultrasound and shock waves that liquids and gases should be at least conventionally divided into "Stokes" and "non-Stokes" fluids. The desire to analyze the nature of bulk viscosity from the viewpoint of mechanics at that level of available experimental data and theoretical concepts can be traced in the materials of the discussion [19], in which many prominent scientists participated.

The main conclusion of the majority of researchers [19] can be formulated as follows: it is necessary to pass to independent coefficients of shear and second viscosities, as proposed much earlier by Poisson, and to avoid considering the Stokes relation as universal. It can be noted that under considerations of bulk (second) viscosity in gases the concept "relaxation" has appeared. Monatomic gases, in contrast to molecular gases, were classified as "Stokes" gases. It is true that was primarily made on the basis of ultrasound experiments with noble gases in which absorption completely coincided with that predicted by the Kirchhoff theory, as opposed to molecular gases. From the viewpoint of the kinetic theory this idea was confirmed by using the definitions of the stress tensor and temperature as a measure of the energy of translational motion of molecules, which were introduced for a monatomic (structureless) gas much earlier by Maxwell. With the use of integrals over molecular velocities they are presented as follows [5, 22] (cf. (1.8)):

$$P_{ij} = m \int c_i c_j \, d\mathbf{c},$$

$$3\rho RT = \left(P_{11} + P_{22} + P_{33}\right) = m \int \left(c_1^2 + c_2^2 + c_3^2\right) d\mathbf{c}. \tag{1.39}$$

Because the hydrostatic (thermodynamic) pressure in the gas is expressed via the equation of state

$$p = \rho R T,$$

it directly follows from Eq. (1.8) that $p^m = p$ and the Stokes relation is valid for monatomic gases. It can be noted that these relations are still presented in some references as the "proof" of the Stokes relation (which can be interpreted as the absence of bulk viscosity) for monatomic gases within the framework of the kinetic theory. We have to pay tribute to the discussion [19, 21], where the participants clarified that it is not a proof at all; it is rather a successful Maxwell's definition. Nevertheless it is rather difficult to provide a reasonable alternative to this relation.

At the same time the materials of that discussion show how slowly the results of the kinetic theory penetrated into the minds of specialists in fluid mechanics. Actually, the modern kinetic theory of a moderately dense monatomic gas, whose validity was confirmed by the closeness of the theoretically calculated and experimental values of the molecular transport coefficients for noble gases, was developed by Chapman and Enskog in the 1920s(!). In particular the stress tensor in a monatomic gas is expressed in the first approximation of the Chapman–Enskog method via the nondivergent strain rate tensor:

$$P_{ij} = p\,\delta_{ij} - \eta\left(\frac{\partial u_i}{\partial x_j} + \frac{\partial u_j}{\partial x_i} - \frac{2}{3}\frac{\partial u_i}{\partial x_i}\right).$$

This result is a direct proof of the absence of bulk viscosity in monatomic gases. It would be by a tautology to say that the divergent component is presented here (cf. Eq. (1.19)) only by the hydrostatic pressure and hence the Stokes relation is satisfied. The Chapman–Enskog theory was mentioned in [19, 21], but not much attention was given. One possible explanation for this fact expressed in [21] was an absence of proof of the convergence of the asymptotic Chapman–Enskog series. However, it is equivalent to rejection of Prandtl's boundary layer theory for the reason that the convergence of the expansion in $\mathrm{Re}^{-1/2}$ was not proved!

Nevertheless, the kinetic theory of molecular gases was at the stage of primary evolution at that time, and results of Wang Chang and Uhlenbeck on informal derivation of the kinetic equation and hydrodynamic equations were not yet obtained. Therefore, palliative attempts based on redefining the Maxwell relations (1.39) for a gas with an internal structure were made in the kinetic theory (see, e.g., [23] and Remark B in [22]) and in mechanics [19, 21] to justify the bulk viscosity effect observed in experiments on ultrasound absorption, in contrast to the direct approach used in [9, 10]. Those attempts were expected to ensure different values of the mean and hydrostatic pressures and hence a nonzero divergent term in Eq. (1.38) representing the second viscosity. The analysis [21] of the influence of various versions of determining the stress tensor and the mean energy in the gas on Eq. (1.38) resulted in the following statement: *"if the hydrostatic (thermodynamic) pressure p and the mean pressure p^m are functions of the state of the system, then the Stokes relation is satisfied for a compressible medium with the stress tensor* (1.35) *and both pressures are identical."*

Indeed, if both pressures are functions of the gas state only, then it follows from Eq. (1.38) that their difference should remain unchanged if the velocity divergence changes. However this happens if and only if $2\eta + 3\lambda = 0$ and $p^m = p$. In fact, the thermodynamic pressure is always determined as a function of state, and the only possible alternative is for p^m. Obviously, in the case with the second viscosity, the mean pressure is no longer a function of the thermodynamic state of the gas only. This statement assisted little in the progress in understanding the second viscosity concept, but it will be useful to conclude our historical outline.

In the kinetic theory of gases the problem of bulk viscosity was exhausted when adequate kinetic equations for polyatomic gases and their asymptotic solutions by the generalized Chapman–Enskog method [4–6] were obtained in the 1960–70s. The corresponding systems of hydrodynamic equations in various approximations were given in Sect. 1.2. The presence of internal energy in polyatomic gas molecules uniquely defines the emergence of divergent terms of the form $G_j \operatorname{div} \mathbf{u}$ in the structure of the first-order approximation corrections (1.17) and (1.28). A direct consequence is the emergence of a divergent term with the bulk viscosity coefficient in the stress tensors (1.19) and (1.29). The known limits of applicability of the developed theory, the consistent character of its conclusions, and good-quality comparability with experimental data eliminate previously existing uncertainty with respect to bulk viscosity.

Nevertheless attempts are made from time to time to cast doubts on the physical reality of bulk viscosity again. A typical example is the publication [18] with a provocative title: “Bulk viscosity as relaxation parameter: Fact or Fiction?”. We have to confess that it is seen from the paper that the authors are well aware of the derivation of the relaxing gas Eqs. (1.21) and (1.31) by the generalized Chapman–Enskog method. For this reason it is hard to understand their attempt to “correct” the known expression for bulk viscosity (1.20):

$$\eta_b = \left(\frac{\mathrm{k}\, c_V^{int}}{m\, c_V^2}\right) p\, \tau_{int}$$

as applied to far ultrasound with frequencies $\omega\tau_{int} \gg 1$. As a result, staying actually within the framework of the same approach, the authors obtained the expression

$$\eta_b = \left[\frac{\mathrm{k}\, c_V^{int}}{m\, c_V^2(1 + \omega^2 \tau_p \tau_v)}\right] p\, \tau_{int},$$

where $\tau_p \sim \tau_v \sim \tau_{int}$ are certain characteristic times. It becomes obvious from here that the result violates the logic of the Chapman–Enskog method, which is based on the expansion in the small parameter

$$\varepsilon = \tau_{int}/\tau_f < 1,$$

because the inverse(!) inequality is valid in this case:

$$\tau_f = 1/\omega \ll \tau_{int}.$$

In this case not only the results of the Chapman–Enskog method, but also the original kinetic equations are inapplicable. Therefore, it is necessary to pass to another level of the flow description in such problems, e.g., using molecular dynamics. Actually the authors come to this conclusion despite the provocative title and the abstract of their paper.

To conclude, we would like to mention one more clearly expressed effect of bulk viscosity manifestation. Returning to Eq. (1.38), we can present it as

$$p^m = p - \eta_b \operatorname{div} \mathbf{u}.$$

It is seen from here that, in accordance with the above-given statement from [21], the mean pressure p^m at $\eta_b \neq 0$ is not a function of the thermodynamic state of the molecular gas only, but it also depends on the local structure of the flow. In divergent flows where $\operatorname{div} \mathbf{u} > 0$ (see the continuity equation) the mean pressure (normal stress) is smaller than the thermodynamic pressure, whereas the reverse situation is observed in convergent flows where $\operatorname{div} \mathbf{u} < 0$.

1.3.2 Rotational Relaxation

Rotational degrees of freedom are excited already at temperatures only several degrees higher than absolute zero [24, 25], where the thermal motion of molecules is hardly manifested at all. The reason is that RT-exchange between translational and rotational degrees of freedom for most diatomic gases (except for molecular hydrogen) occurs rather easily. Therefore, the process of rotational relaxation is inherent in almost all molecular gas flows, though its manifestation is not always obvious and is often disguised by more intense processes.

Vibrational modes in molecular gas flows with moderate supersonic velocities, which are characterized by stagnation temperatures T_* of the order of 1000 K, are weakly excited and the fraction of internal energy of these modes is negligibly small. A natural gas-dynamic model that takes into account dissipative effects is the Navier–Stokes equations (1.21), where relaxation is expressed via bulk viscosity.

In this case the general Eq. (1.20) for bulk viscosity is transformed to

$$\eta_b = \left(\frac{\mathrm{k}\, \gamma_r}{m\, c_V} \right) p\, \tau_{rt}, \tag{1.40}$$

where $p = n\mathrm{k}T$ is the hydrostatic pressure of the gas, c_V is the total specific heat, γ_r is the fraction of internal energy of rotational degrees of freedom of molecules, and τ_{rt} is the time of rotational relaxation. For linear diatomic molecules, assuming equipartition over the degrees of freedom, we have $c_V = 5\mathrm{k}/(2m)$ and $c_{V,r} = \mathrm{k}/m$, $\gamma_r = c_{V,r}/c_V = 2/5$. Thus, to find the bulk viscosity coefficient for a particular gas

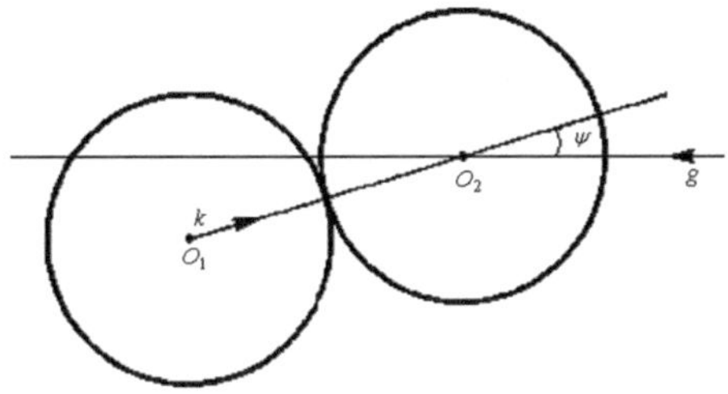

Fig. 1.1 Model of rough spheres

one has to know the corresponding time τ_{rt}. It is often estimated by the formula [6] (see also [4])

$$\tau_{rt} = Z_r(T)\tau_\eta. \tag{1.41}$$

Here, T is the flow temperature, τ_η is the mean time between elastic collisions of gas molecules, and $Z_r(T)$ is the coefficient of energy exchange between rotational and translational degrees of freedom of molecules, which defines the mean number of molecular collisions necessary for relaxation of rotational energy of molecules to the equilibrium state.

Unfortunately, there are no systematic data on bulk viscosity coefficients in modern literature. We can rather state that these data are scarce and often contradictory. The values of $Z_r(T)$ and τ_η in Eq. (1.41) are often estimated in the kinetic theory by various approximate approaches based on simple models of molecular interaction [6, 22, 26–28]. It seems reasonable to give here some formulae for these variables and also experimental data that we managed to find during our study. If necessary these reference materials will allow one to find a fairly justified range of variation of the basic parameters of rotational relaxation.

There are no quantum effects in rotational relaxation in the temperature range of interest for us [6]. Therefore reasonable estimates of rotational relaxation times are obtained from the classical mechanical models of molecular interaction. In first turn, the "solid" molecules are used, such as rough spheres [22, 26–28], loaded spheres [29–31], and spherical cylinders [32–34]. Here we present the results for the model of rough spheres, which is widely used because it allows one to obtain simple analytical expressions for the kinetic characteristics of rotational relaxation. In addition to τ_{rt} this model gives explicit formulae for coefficients of shear viscosity, thermal conductivity, and diffusion [6]. Moreover, as compared to other similar models, the model of rough spheres provides the most adequate description of the specific feature of inelastic collisions: RT-transition of translational to rotational energy. In the model of rough spheres a molecule is an impermeable sphere of diameter d and mass m, which is spherically-symmetrically distributed with respect to the geometric center of the molecule located at the point O (see Fig. 1.1). The roughness of the spheres is understood in the following meaning: the relative velocity of the spheres at the point of their contact changes its sign to the opposite one at the instant of the collision. The collision dynamics is characterized by the parameter $K = 4I/\left(md^2\right)$, where I is the moment of inertia of the molecule with respect to the axis of rotation passing through its center. The value of the parameter K depends on the distribution of the molecular mass over the sphere volume. In particular, when $K = 0$ the entire mass of the molecule is concentrated at its center (material point) and when $K = 2/3$ the molecular mass is distributed over the surface, which is the maximum possible value. Thus, the values of this parameter are in the interval $0 < K \leq 2/3$.

The rotational relaxation time in the model of rough spheres is calculated by the formula [6]

$$\tau_{rt} = \frac{3(1+K)^2}{16K\, n\, d^2}\sqrt{\frac{m}{\pi \mathrm{k} T}}. \tag{1.42}$$

In accordance with Eq. (1.40), the bulk viscosity coefficient is

$$\eta_b = \frac{3}{100}\frac{(1+K)^2}{K}\frac{1}{d^2}\sqrt{\frac{mkT}{\pi}}. \tag{1.43}$$

The time between elastic collision, which is of the order of the mean free path time, is

$$\tau_\eta = \frac{3}{8}(n\pi d^2)^{-1}\sqrt{\frac{\pi m}{kT}}. \tag{1.44}$$

As a result the coefficient of energy exchange between translational and rotational degrees of freedom of molecules (1.41) is expressed as

$$Z_r = \frac{12}{5\pi}\frac{(1+K)^2}{K}. \tag{1.45}$$

It is seen that this coefficient is independent of the flow temperature T. This is a drawback of this model because it contradicts ultrasonic measurements of the exchange coefficient Z_r in [24, 25, 35, 36], where Z_r was observed to increase monotonically with increasing gas temperature T. It also follows from equality (1.45) that the energy exchange coefficient at $K \to 0$ (the entire mass of the molecule is concentrated at its center) tends to infinity as $Z_r \sim 1/K$. In the case with $K = 2/3$ (the entire mass of the molecule is distributed over its surface), the exchange coefficient takes the minimum value equal to $Z_r = 10/\pi \approx 3.183$.

The shear viscosity η is defined as

$$\eta = \frac{15}{8\, d^2}\frac{(1+K)^2}{(6+13K)}\sqrt{\frac{m\mathrm{k}T}{\pi}}. \tag{1.46}$$

Using Eqs. (1.43) and (1.46), we obtain the ratio of bulk viscosity to shear viscosity

$$\alpha_1 = \frac{\eta_b}{\eta} = \frac{2}{125}\frac{6+13K}{K}, \quad 0 < K \leq 2/3, \tag{1.47}$$

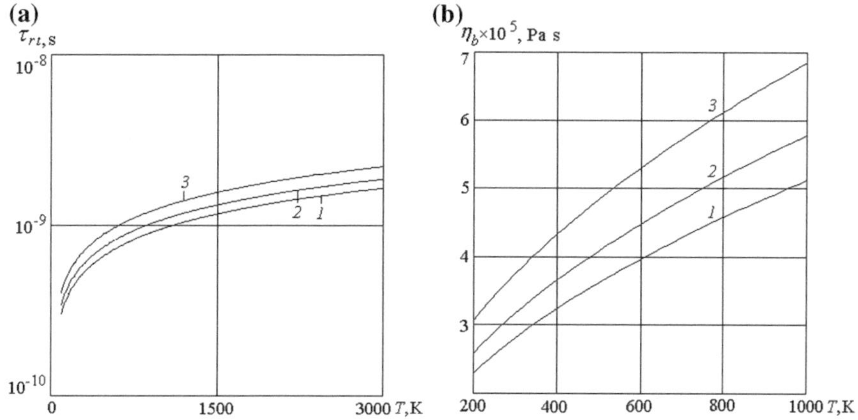

Fig. 1.2 Temperature dependencies of the rotational relaxation time τ_{rt} and bulk viscosity η_b calculated in the rough sphere approximation for the gas pressure $p = 1$ atm and $K = 2/5$. **a** shows dependencies $\tau_{rt}(T)$. **b** shows dependencies $\eta_b(T)$. Curves *1–3* show the data for nitrogen, oxygen, and carbon monoxide, respectively

which shows that ratio (1.47) tends to infinity in accordance with the expression $\alpha_1 \sim 1/K$ as $K \to 0$ and takes its minimum value equal to $\alpha_1 \approx 0.352$ at $K = 2/3$.

Figure 1.2 shows the temperature dependencies of the rotational relaxation time τ_{rt} and bulk viscosity η_b calculated for nitrogen, oxygen, and carbon monoxide in the rough sphere approximation for the parameter $K = 2/5$ and gas pressure $p = 1$ atm. At $K = 2/5$ the entire mass of the molecule is uniformly distributed over the sphere volume.

Results in reasonable agreement with experimental data on the temperature dependence $Z_r(T)$ are provided by Parker's formula [37], where a modified Morse potential is used as a model of molecular interaction.

The analytical expression for $Z_r(T)$ has the form [37, 38]

$$Z_r(T) = Z_\infty \left[1 + \frac{\pi^{3/2}}{2}\sqrt{\frac{\theta}{T}} + \left(2 + \frac{\pi^2}{2}\right)\frac{\theta}{T} + \left(\frac{\pi\theta}{T}\right)^{3/2} \right]^{-1}, \qquad (1.48)$$

where θ is the parameter of the interaction potential. There are recommendations of choosing the value of this parameter for some gases in the literature. In particular it was proposed in the calculations [39, 40] to use the value $\theta = 91.5$ K for nitrogen N_2 and $\theta = 88$ K for oxygen O_2 and carbon monoxide CO. The coefficient Z_∞ is used as an empirical fitting parameter whose value is determined from comparisons of theoretical and experimental values of $Z_r(T)$ for a certain value of the gas temperature T.

Figure 1.3a–c show the calculated temperature dependencies $Z_r(T)$ for nitrogen N_2, oxygen O_2, and carbon monoxide CO [40]. The points are the experimental values of the factor Z_r obtained by processing experimental data on ultrasound absorption in

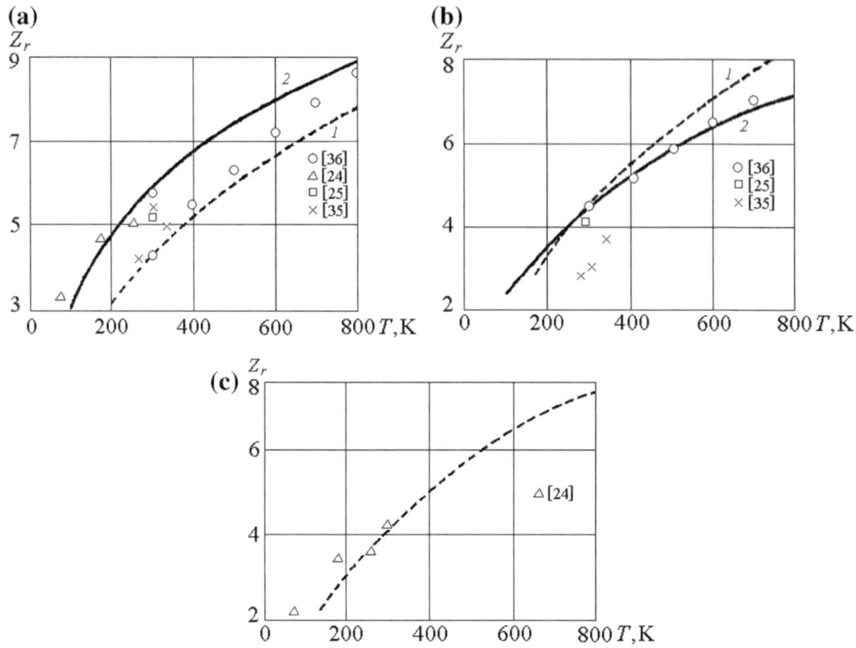

Fig. 1.3 Dependencies $Z_r(T)$ for nitrogen N_2, oxygen O_2, and carbon monoxide CO (see the comments to this figure in the text). **a** shows $Z_r(T)$ for N_2. **b** shows $Z_r(T)$ for O_2. **c** shows $Z_r(T)$ for CO

these gases. The experimental values of Z_r for the gas temperature $T = 300$ K were taken from [24, 25, 35, 36] and have the following values for the diatomic gases considered here: $Z_r = 4.3$ for N_2, $Z_r = 4.5$ for O_2, and $Z_r = 4.1$ for CO. The values of the parameter Z_∞ were determined on the basis of coincidence of theoretical and experimental values of the factor $Z_r(T)$ at the gas temperature $T = 300$ K. It was found that $Z_\infty = 24.04$ for nitrogen, $Z_\infty = 24.42$ for oxygen, and $Z_\infty = 22.25$ for carbon monoxide.

Curves *1* in Fig. 1.3a–c are obtained in calculations by Eq. (1.48), and curves *2* are the numerical calculations [40]. It is seen that the theoretical curves calculated by the classical method adequately describe the observed temperature dependence $Z_r(T)$ for the diatomic molecules considered in this study. Additional experimental data on ultrasound measurements of the factor $Z_r(T)$ at room temperature can be found in [3].

Formula (1.48) and relations (1.40) and (1.41) were used to calculate the temperature dependencies of bulk viscosity $\eta_b(T)$ and rotational relaxation time $\tau_{rt}(T)$ for nitrogen, oxygen, and carbon monoxide. The time between molecular collisions τ_η was calculated by formula (1.44), where the particle density $n = p/\,(\mathrm{k}T)$ was taken to be equal to the number of gas molecules in a unit volume at the pressure $p = 1$ atm and temperature $T = 273$ K and remained unchanged during the calculation. The calculated values of rotational relaxation time $\tau_{rt}(T)$ and bulk viscosity $\eta_b(T)$ for nitrogen, oxygen, and carbon monoxide are shown in Fig. 1.4.

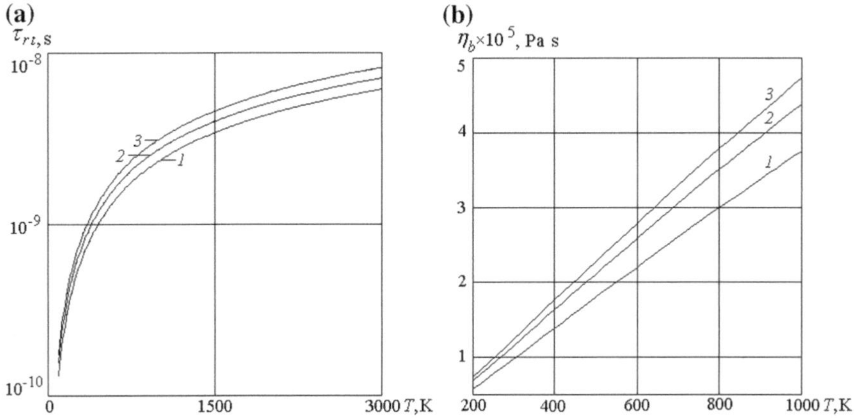

Fig. 1.4 Temperature dependencies of the rotational relaxation time τ_{rt} and bulk viscosity η_b. **a** shows dependencies $\tau_{rt}(T)$. **b** shows dependencies $\eta_b(T)$. Curves *1–3* show the results for nitrogen, oxygen, and carbon monoxide, respectively

A comparison of the plots of the functions $\eta_b(T)$ and $\tau_{rt}(T)$ in Figs. 1.4 and 1.2 leads to the following conclusion. The values of η_b obtained in the rough sphere approximation for the flow temperature T in the interval from 200 to 1000 K are higher by an order of magnitude on the average than the bulk viscosity values calculated by Parker's model (see Figs. 1.2b and 1.4b). For the rotational relaxation times however the reverse relationship is observed (see Figs. 1.2a and 1.4a).

Modeling of relaxing gas flows by the Navier–Stokes equations (1.21) or using two-temperature gas-dynamic Eq. (1.31) in the dimensionless form usually involves the ratio $\alpha_1(T) = \eta_b(T)/\eta(T)$ of the bulk and shear viscosities.

Figure 1.5 shows the calculated dependencies $\alpha_1(T)$ for nitrogen, oxygen, and carbon monoxide. The shear viscosity $\eta(T)$ is determined by the empirical Sutherland's formula [41–43]

$$\frac{\eta(T)}{\eta_0} = \frac{293 + C}{T + C}\left(\frac{T}{293}\right)^{3/2},$$

where η_0 is the shear viscosity at the flow temperature $T = 293$ K and C is the empirical constant in Sutherland's law. The values of η_0 and C for nitrogen, oxygen, and carbon monoxide were taken from [41–43]. It follows from Fig. 1.5 that the ratio $\alpha_1 = \eta_b/\eta$ increases with increasing temperature. The estimates for molecular gas flows with moderate parameters within the limits of applicability of the Navier–Stokes model yield $\alpha_1 < 2$.

It is also reasonable to give some experimental data on bulk viscosity values for molecular gases. The earlier results are based on ultrasound absorption measurements. In the linear acoustic approximation the absorption coefficient is determined by the formula [9, 10, 41–43]

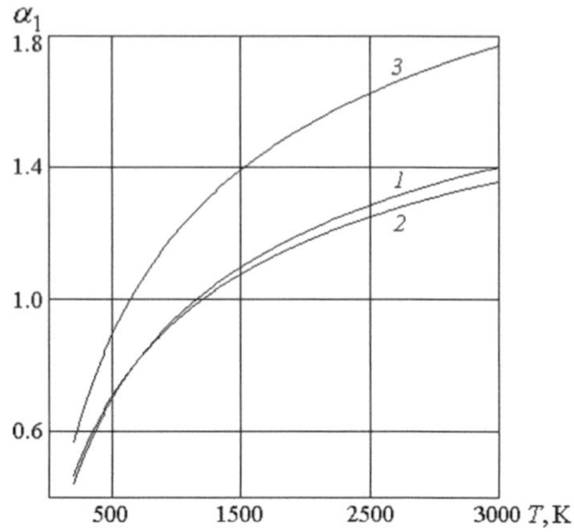

Fig. 1.5 Dependencies $\alpha_1(T)$. Curves *1*–*3* show the results for nitrogen, oxygen, and carbon monoxide, respectively

Table 1.1 Bulk viscosity of some gases under standard conditions $T = 273$ K and $p = 10^5$ Pa

Gas	N_2	CO	Air	CO_2	H_2
$\eta \times 10^5$, Pa s	1.750	1.750	1.820	1.460	0.880
$\eta_b^{(s)} \times 10^5$, Pa s	0.348	12.274	1.178	54.015	34.570
$\alpha_1^{(s)} = \eta_b^{(s)}/\eta$	0.199	7.014	0.647	39.997	39.280
$\eta_b^{(r)} \times 10^5$, Pa s	0.966	1.231	–	–	–
$\alpha_1^{(r)} = \eta_b^{(r)}/\eta$	0.552	0.703	–	–	–

$$\kappa = \frac{2\pi\omega^2}{\rho c_s^2}\left[\frac{4\eta}{3} + \eta_b^{(s)} + \frac{(\gamma-1)\lambda}{c_p}\right], \tag{1.49}$$

where λ is the thermal conductivity, γ is the ratio of specific heats, c_s is the velocity of sound, c_p is the specific heat at constant pressure, and ω is the ultrasonic wave frequency. The data on the absorption coefficient κ were taken from [41–43]. The values of bulk viscosity $\eta_b^{(s)}$ calculated by Eq. (1.49) for the gas pressure $p = 10^5$ Pa and gas temperature $T = 273$ K are summarized in Table 1.1.

More recent data on the bulk viscosity coefficient are obtained from electron-optical measurements of the rotational relaxation times τ_{rt} or exchange coefficients Z_r in shock waves and underexpanded jets. Some values of $\eta_b^{(r)}$ calculated by formulae (1.40) and (1.41) with the use of experimental data taken from some papers cited in [41–43] are also listed in Table 1.1.

These results agree well with bulk viscosity values predicted by various kinetic models including Parker's model (see Fig. 1.4) and with more recent experimental data of other researchers [41–43]. The difference in the values of $\eta_b^{(s)}$ and $\eta_b^{(r)}$, approximately by an order of magnitude, for carbon monoxide CO may be cansed

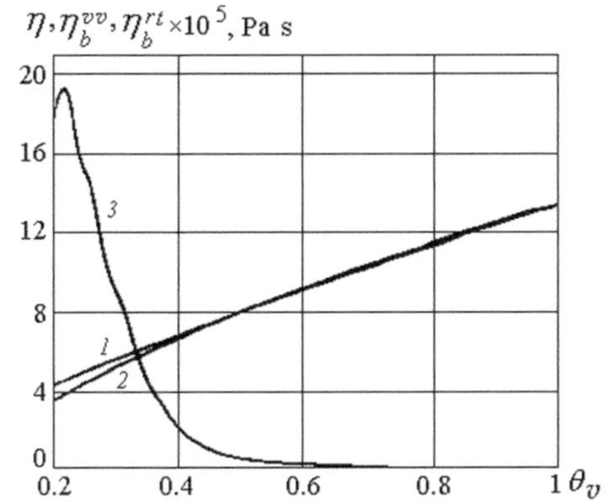

Fig. 1.6 Bulk η_b^{rt}, η_b^{vv} and shear η viscosities versus the degree of excitation of vibrational modes of molecules θ_v. Curve *1* shows the dependence for the shear viscosity η. Curves *2* and *3* show the results for the bulk viscosities η_b^{rt} and η_b^{vv}, respectively

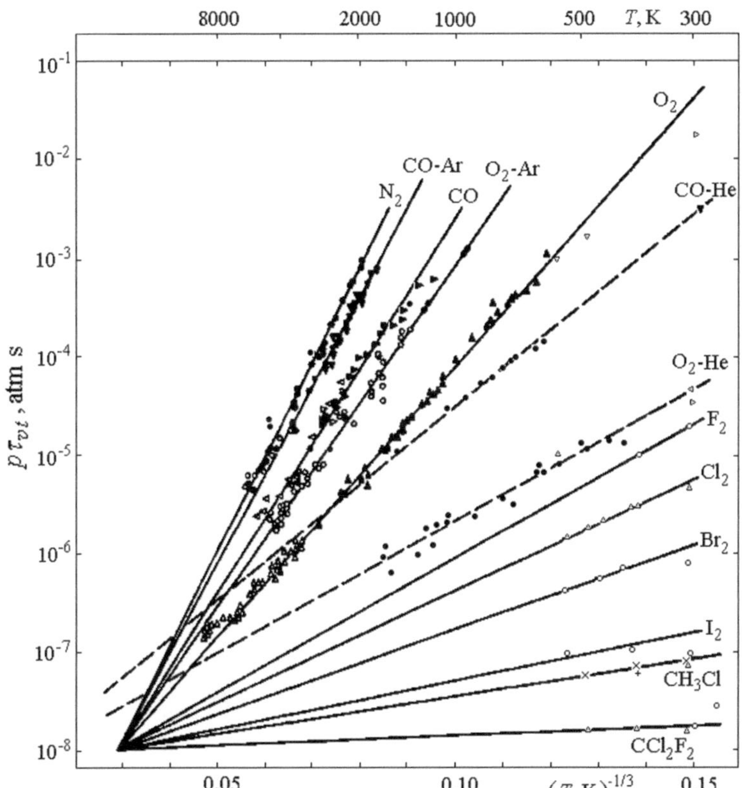

Fig. 1.7 Vibrational relaxation times

by particular sensitivity of acoustic measurements to the presence of even small fractions of impurities in the gas, especially water vapor [41–43]. At the same time it follows from the comparison of the values of $\eta_b^{(r)}$ that bulk viscosity of carbon monoxide at the present values of gas pressure and temperature is higher than that of nitrogen only by one third.

1.3.3 Vibrational Relaxation

For calculating vibrationally excited gas flows by equations of two-temperature aerodynamics (1.31) it is necessary to know the range of variation of vibrational relaxation times τ_{vt} and the contribution of vibrational relaxation to the bulk viscosity coefficient.

In the hierarchy of kinetic times (1.2) considered above, where vibrational relaxation passes through the stage of reaching the Boltzmann distribution over vibrational energy levels and the two-temperature model (1.31) is applicable, the bulk viscosity coefficient is presented as the sum of two terms

$$\eta_b = \eta_b^{rt} + \eta_b^{vv}.$$

The first term is related to rotational relaxation and can be calculated by Eq. (1.41). It would be formally possible to use a similar formula for the second term associated with resonance VV-exchange by vibrational quanta with the characteristic time τ_{vv}. However, there are no necessary data (in particular on the relaxation times τ_{vv}) in publications because only the parameter τ_{vt} is measured in experiments. It was demonstrated [4] that $\eta_b^{vv} = 0$ in the model of harmonic oscillations. In the case of intense excitation of vibrational modes however the addition to η_b due to anharmonicity of oscillations may become significant.

It follows from the results [44–46] that the coefficient η_b^{vv} for the vibrational nonequilibrium level $\theta_v = T/T_v \leq 1/4$, which is reached for example in supersonic underexpanded jets at Mach numbers M = 4–5, may exceed the value of η_b^{rt} by a factor of 2–5.[1] Unfortunately no systematic calculations of this kind have been performed until recently. To give the reader a general idea, we present the values of η_b^{vv} calculated in [4] for molecular nitrogen. In addition Fig. 1.6 shows the plots for η_b^{rt} and shear viscosity η. It is seen that this effect can be significant in the case of laser pumping of the vibrational mode in a comparatively cold gas flow.

A sufficient approximation for the vibrational relaxation time τ_{vt} in aerodynamic problems is provided by the Landau–Teller theory [1]. In addition to the vibrational energy relaxation Eq. (1.26), this theory offers the linear dependence

---

[1]Here and in Fig. 1.6 the vibrational nonequilibrium is characterized by the temperature T_v, which is associated in a certain approximation with the temperature of the first vibrational level T_1 used in [4].

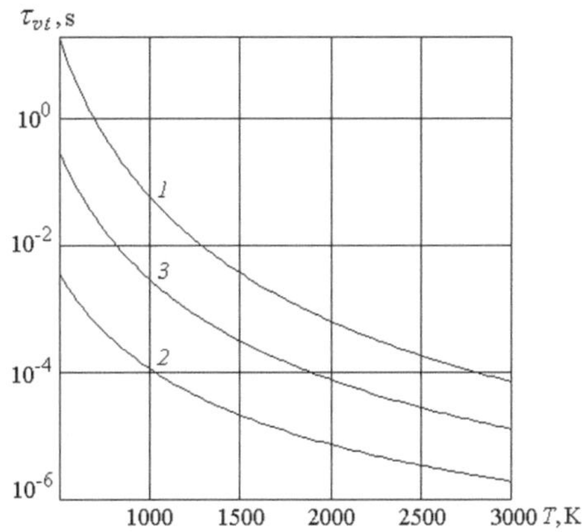

Fig. 1.8 Dependencies $\tau_{vt}(T)$ for nitrogen (*1*), oxygen (*2*), and carbon monoxide (*3*)

$$\ln(p\,\tau_{vt}) \sim T^{-1/3},$$

which is valid for a number of gases including diatomic gases. The corresponding experimental data systematized in [47] are plotted in Fig. 1.7. References to original sources of experimental data can be found in [47].

The correlating curves in Fig. 1.7 are approximated by the semiempirical Millikan-White formula

$$\ln(p\,\tau_{vt}) = A\left(T^{-1/3} - 0.015\,\mu^{1/4}\right) - 18.420. \tag{1.50}$$

The dimensional variables in Eq. (1.50) are expressed in the following units: VT-relaxation time τ_{vt} in seconds, gas pressure p in atmospheres, gas temperature T in Kelvin degrees, and collision reduced mass of colliding molecules μ in atomic units. The coefficient A depends on molecular constants. Tables with the corresponding values of A are available in [47] for the gases in Fig. 1.7.

Figure 1.8 shows the temperature dependencies of VT-relaxation times $\tau_{vt}(T)$ calculated by Eq. (1.50) for nitrogen N_2, oxygen O_2, and carbon monoxide CO. The empirical constants A for these diatomic gases taken from [47] are $A = 220$ for N_2, $A = 129$ for O_2, and $A = 175$ for CO.

It follows from Fig. 1.8 that the VT-relaxation times τ_{vt} for the gas temperature T ranging from 500 to 3000 K and gas pressure $p = 1$ atm vary approximately from 10 to 10^{-6} s and decrease with increasing T.

1.4 Absorption of Acoustic Waves in the Relaxation Process

Disturbances whose instability and dissipation in shear flows are studied in subsequent chapters are acoustic waves in accordance with the definition [48]. Therefore, to understand these processes one has to know the mechanisms of absorption of acoustic waves in flow on the background of the relaxation process in gas. The material in this section basically follows [14], where more details can be found.

The system of acoustic equations in a relaxing gas can be obtained by means of linearization of system (1.31) with respect to the equilibrium quiescent state with the parameters ρ_0, p_0, and E_{v0} where viscous diffusion and heat conduction are neglected. In a spatially one-dimensional case it has the form

$$\frac{\partial \rho}{\partial t} + \rho_0 \frac{\partial u}{\partial x} = 0, \quad \rho_0 \frac{\partial u}{\partial t} + \rho_0 \frac{\partial p}{\partial x} = 0, \quad \rho_0 \frac{\partial T}{\partial t} - \frac{\partial p}{\partial t} = 0$$

$$\frac{dE_v}{dt} = \frac{E_{v0}(T) - E_v}{\tau_0}. \tag{1.51}$$

System (1.51) is closed by the linearized equation of state. The primes at small perturbations with respect to which the equations are written are omitted (see [14]).

We introduce a potential function $\varphi(t, x)$ identically satisfying the momentum equation if we assume that

$$p = -\rho_0 \frac{\partial \varphi}{\partial t}, \quad u = \frac{\partial \varphi}{\partial x}. \tag{1.52}$$

The system of acoustic equations reduces to the third-order scalar equation

$$\tau_0 \left(\frac{1}{a_f^2} \varphi_{tt} - \varphi_{xx} \right)_t + \left(\frac{1}{a_e^2} \varphi_{tt} - \varphi_{xx} \right) = 0, \tag{1.53}$$

where the subscripts mean differentiation of the potential function over the corresponding variables. This system contains two velocities of acoustic waves: a_e and a_f. The "equilibrium" velocity a_e corresponds to the case with no relaxation process ($\tau_0 = 0$). The opposite case ($\tau_0 \to \infty$) of a slow process is described by the "frozen" velocity of sound a_f.

In the equilibrium case the potential function obviously satisfies the classical wave equation

$$\varphi_{tt} - a_e^2 \varphi_{xx} = 0,$$

whose general solution is written [49] as a superposition of two plane waves

$$\varphi(x, t) = f(x - a_e t) + g(x + a_e t),$$

moving in opposite directions with an identical velocity a_e with no changes in the profile shape. Therefore, it should be emphasized that the transition to Eq. (1.53) and the results of its analysis are determined exclusively by the nonequilibrium relaxation process.

For the analysis it is sufficient to consider the evolution of a small-amplitude harmonic perturbation introduced into the gas by using the boundary condition on the perturbation velocity $u(o, t) = \varepsilon\omega \cos \omega t$. The sought velocity of the perturbation satisfies the equation

$$\tau\left(\frac{1}{a_f^2}u_{tt} - u_{xx}\right)_t + \left(\frac{1}{a_e^2}u_{tt} - u_{xx}\right) = 0,$$

which is obtained with allowance for Eq. (1.52) by differentiation of Eq. (1.53) with respect to x. Its solution has the form

$$u(x, t) = \varepsilon\omega \exp\left[-\frac{\omega\delta}{a_f}x\right]\cos\left[\omega\left(t - \frac{\lambda}{a_f}\right)\right]. \tag{1.54}$$

Here

$$\begin{Bmatrix}\delta\\ \lambda\end{Bmatrix} = \left\{\frac{1}{2\left(1+\omega^2\tau^2\right)}\left[\mp(b+\omega^2\tau^2) + \sqrt{(1+\omega^2\tau^2)\left(b^2+\omega^2\tau^2\right)}\right]\right\}^{1/2},$$

$$b = a_f^2/a_e^2.$$

The upper sign refers to δ, and the lower sign refers to λ. As is seen from Eq. (1.54), a harmonic perturbation propagates in a relaxing gas with a phase velocity $V = a_f/\lambda$. Its amplitude exponentially decays with a factor $\Delta = \omega\delta/a_f$ per unit length. Both variables, λ and δ, and hence the phase velocity and damping depend on the frequency and relaxation time. The analysis of the above-given relations shows that the phase velocity and damping are enhanced with increasing frequency. A simple estimate of damping Δ can be obtained by the semiempirical relation [50]

$$\Delta = C\frac{\omega\tau}{1+\omega^2\tau^2},$$

where the constant C depends only on the type of gas. It follows from here that the maximum damping is reached at the "relaxation" frequency $\omega_m = 1/\tau$. Thus, absorption of the perturbation energy in the relaxing gas has a resonance character.

Equations (1.52) and (1.53) yield the expression for the pressure disturbance

$$p = \frac{\varepsilon\omega\rho_0 a_f}{\sqrt{\delta^2+\lambda^2}}\exp\left[-\frac{\omega\delta}{a_f}x\right]\cos\left[\omega\left(t + t_p - \frac{\lambda}{a_f}\right)\right].$$

It is seen from a comparison with Eq. (1.54) that the pressure disturbance propagates with the same velocity V and exponentially decays with the same factor Δ, but its front at a given point x leading the velocity disturbance in phase on angle

$$\psi = \omega t_p = \arctan\left(\frac{\delta}{\lambda}\right).$$

It can be shown [50] that the maximum phase shift occurs at the same "relaxation" frequency ω_m as the maximum damping. It follows from the continuity equation in system (1.51) that the same phase shift is observed between the density and pressure perturbations. Physically, such a discrepancy is explained by a finite time of relaxation of vibrational modes (compression waves) as compared to the instantaneous response of translational degrees of freedom to the perturbation, whose momentum determines the pressure wave.

By introducing a perturbation of the specific volume $v = 1/\rho$ the behavior of the harmonic perturbation wave parameters at a given point x can be presented in the pv-diagram as a closed thermodynamic cycle formed by two pairs of equidistant polytropic curves [50]. The area bounded by these curves defines the energy absorbed during one oscillation. In turn the absorbed (or dissipated to heat) energy is associated with an increase in entropy in the relaxation process, which is mentioned in Sect. 3.2.3.

As a concluding remark, we should mention the character of propagation of an arbitrary perturbation. Such a perturbation can be presented as the Fourier series (integral) over the harmonic components. As higher frequencies propagate faster and are absorbed more intensely, an arbitrary perturbation in a relaxing gas is always deformed and selectively decays.

References

1. Gordiets, B.F., Osipov, A.I., Shelepin, L.A.: Kinetic Processes in Gases and Molecular Lasers. Gordon and Breach, New York (1987)
2. Losev, S.A.: Gasdynamic Laser. Springer, Berlin (1981)
3. Stupochenko, E.V., Losev, S.A., Osipov, A.I.: Relaxation Processes in Shock Waves. Nauka, Moscow (1965) (in Russian)
4. Nagnibeda, E.A., Kustova, E.V.: Non-equilibrium Reacting Gas Flows. Kinetic Theory of Transport and Relaxation Processes. Springer, Berlin (2009)
5. Ferziger, J.H., Kaper, H.G.: Mathematical Theory of Transport Processes in Gases. North Holland Publishing Company, Amsterdam (1972)
6. Zhdanov, V.M., Aliyevskii, M.Ya.: Transfer and Relaxation Processes in Molecular Gases. Nauka, Moscow (1989) (in Russian)
7. Kogan, M.N.: Rarefied Gas Dynamics. Plenum Press, New York (1969)
8. Zeldovich, Ya.B., Raizer, Yu.P.: Physics of Shock Waves and High-Temperature Hydrodynamic Phenomena. Academic Press, New York (1967)
9. Leontovoich, M.A.: Remarks on the theory of sound attenuation in gases. Zh. Éksp. Teor. Fiz. **6**, 561–576 (1936) (in Russian)
10. Leontovoich, M.A.: Some questions in the theory of sound absorption polyatomic gases. Izv. Akad. Nauk SSSR. Ser. Fizika **5**, 633–636 (1936) (in Russian)
11. Schwartz, R.N., Slawsky, Z.I., Herzfeld, K.F.: Calculation of vibrational relaxation times in gases. J. Chem. Phys. **20**, 1591–1599 (1952)
12. Gordiets, B.F., Zhdanok, S.: Analytical theory of vibrational kinetics of anharmonic oscillators. In: Capitelli, M. (ed.) Nonequilibrium Vibrational Kinetics. Series Topics in Current Physics, vol. 39, pp. 47–84. Springer, Berlin (1986)

13. Brun, R., Zappoli, B.: Model equations for a vibrationally relaxing gas. Phys. Fluids **20**, 1441–1448 (1977)
14. Vincenti, W.G., Kruger, Ch.H.: Introduction to Physical Gas Dynamics. Krieger, Huntington (1975)
15. Landau, L., Teller, E.: On the theory of sound dispersion. In: Ter Haar, D. (ed.) Collected Papers of L.D. Landau, pp. 147–153. Pergamon Press, Oxford (1965)
16. Kustova, E.V., Oblapenko, G.V.: Reaction and internal energy relaxation rates in viscous thermochemically non-equilibrium gas flows. Phys. Fluids **27**, 016102 (2015)
17. Kochin, N.E., Kibel, I.A., Rose, N.V.: Theoretical Hydromechanics. Part 2. Fizmatgiz, Moscow (1963) (in Russian)
18. Meador, W.E., Miner, G., Townsend, I.W.: Bulk viscosity as relaxation parameter: fact or fiction? Phys. Fluids **8**, 258–261 (1996)
19. Rosenhead, L., et al.: A discussion on the first and second viscosities. Proc. R. Soc. A. **226**, 1–69 (1954)
20. Lamb, H.: Hydrodynamics. Cambridge University Press, Cambridge (1932)
21. Truesdell, C.: On the viscosity of fluids according to the kinetic theory. Zeitsch. Phys. **131**, 273–289 (1952)
22. Chapman, S., Cowling, T.G.: The Mathematical Theory of Non-uniform Gases. Cambridge University Press, Cambridge (1952)
23. Kohler, M.: Reibung in mabig verdunnten gasen als folge verzogerte einstellung der energie. Zeitsch. Phys. **125**, 715–732 (1949)
24. Prangsma, G.J., Alberga, A.H., Beenakker, J.J.M.: Ultrasonic determination of the volume viscosity of N_2, CO, CH_4, CO_2 between 77 and 3000 K. Physica **64**, 278–288 (1973)
25. Greenspan, M.: Rotational relaxation in nitrogen, oxygen and air. J. Acoust. Soc. Am. **31**, 155–160 (1959)
26. Monchick, L., Yun, K.S., Mason, E.A.: Relaxation effects in the transport properties of rough spheres. J. Chem. Phys. **38**, 1282–1287 (1963)
27. Condiff, D.W., Lu, W.K., Dahler, J.S.: Transport properties of polyatomic fluids. A dilute gas of perfectly rough spheres. J. Chem. Phys. **42**, 3445–3475 (1965)
28. Hirschfelder, J.O., Curtiss, Ch.F., Bird, R.B.: Molecular Theory of Gases and Liquids. Wiley, New York (1954)
29. Jeans, J.H.: The distribution of molecular energy. Philos. Trans. R. Soc. **68**, 175–253 (1901)
30. Dahler, J.S., Satcher, N.F.: Kinetic theory of loaded molecules, I. J. Chem. Phys. **38**, 2363–2382 (1963)
31. Sandler, S.I., Dahler, J.S.: Kinetic theory of loaded molecules, II. J. Chem. Phys. **43**, 1750–1759 (1965)
32. Curtiss, C.F., Dahler, J.S.: Kinetic theory of nonspherical molecules. J. Chem. Phys. **24**, 2425–2441 (1956)
33. Curtiss, C.F., Muckenfuss, C.: Kinetic theory of nonspherical molecules, II. J. Chem. Phys. **26**, 1619–1636 (1957)
34. Curtiss, C.F., Dahler, J.S.: Kinetic theory of nonspherical molecules, V. J. Chem. Phys. **38**, 2352–2362 (1963)
35. Parbrook, H.D., Tempest, W.: Sound absorption in nitrogen and oxygen. Acustica **8**, 345–348 (1958)
36. Carnevale, E.H., Carey, C., Larson, G.: Ultrasonic determination of rotational collision numbers and vibrational relaxation times of polyatomic gasses at high temperatures. J. Chem. Phys. **47**, 2829–2844 (1967)
37. Parker, J.G.: Rotational and vibrational relaxation in diatomic gases. Phys. Fluids **2**, 449–461 (1959)
38. Brau, C.A., Johnkman, K.H.: Classical theory of rotational relaxation in diatomic gases. J. Chem. Phys. **52**, 477–484 (1970)
39. Gerasimov, G.Ya., Makarov, V.N.: Theory of rotational relaxation in a diatomic gas. J. Appl. Mech. Tech. Phys. **16**, 8–12 (1975)

40. Lordi, J.A., Mater, R.E.: Rotational relaxation in nonpolar diatomic gases. Phys. Fluids **13**, 291–308 (1970)
41. Grigor'ev, I.S., Meilikhov, E.Z. (eds.): Physical Quantities, Handbook. Énergoatomizdat, Moscow (1991) (in Russian)
42. Kikoin, I.K. (ed.): Tables of Physical Quantities, Handbook. Atomizdat, Moscow (1976) (in Russian)
43. Kaye, G.W., Laby, T.H.: Tables of Physical and Chemical Constants. Longmans, Green & Co., London (1958)
44. Kustova, E.V., Nagnibeda, E.A.: The influence of non-Boltzman vibrational distribution on thermal conductivity and viscosity. In: Capitelli, M. (ed.) Molecular Physics and Hypersonic Flows, pp. 383–392. Kluwer Academic Publishers, Dordrecht (1996)
45. Kustova, E.V.: Kinetic model of molecular gas dynamics in strongly non-equilibrium conditions. Vestn. S.-Peterb. Univ. Ser. **1**(9), 60–65 (1995) (in Russian)
46. Kustova, E.V., Nagnibeda, E.A.: Strong non-equilibrium effects on specific heats and thermal conductivity of diatomic gas. Chem. Phys. **208**, 313–329 (1996)
47. Millikan, R.C., White, D.R.: Systematics of vibrational relaxation. J. Chem. Phys. **39**, 3209–3213 (1963)
48. Mack, L.M.: On the inviscid acoustic-mode instability of supersonic shear flows. Part I: two-dimensional waves. Theor. Comput. Fluid Dyn. **2**, 97–123 (1990)
49. Korn, G.A., Korn, T.M.: Mathematical Handbook for Scientists and Engineers. McGraw-Hill, New York (1961)
50. Shpakovsky, B.G.: The dispersion and selective absorption of ultrasonic waves in a monatomic gas. Usp. Fiz. Nauk **14**, 283–297 (1934) (in Russian)

Chapter 2
Linear Stability of Inviscid Plane-Parallel Flows of Vibrationally Excited Diatomic Gases

Abstract This chapter is devoted to investigations of linear stability of plane-parallel flows of an inviscid nonheat-conducting vibrationally excited gas. Some classical results of the theory of linear stability of ideal gas flows, such as the first and second Rayleigh's theorems and Howard's theorem, are generalized. An equation of the energy balance of disturbances is derived, which shows that vibrational relaxation generates an additional dissipative factor, which enhances flow stability. Calculations of the most unstable inviscid modes with the maximum growth rates in a free shear layer are described. It is shown that enhancement of excitation of vibrational modes leads to reduction of the growth rates of inviscid disturbances.

Linear stability of plane-parallel shear flows are traditionally studied within the framework of the hydrodynamic stability theory. Such studies for an ideal incompressible fluid were performed in the classical works of Helmholtz, Kelvin, and Rayleigh. Later on, their results were extended to more realistic problems of ideal compressible gas flows and inhomogeneous stratified and conducting fluid flows in fields of various mass forces [1].

This chapter describes the results of studies of linear stability of plane-parallel shear flows of an inviscid non-heat-conducting vibrationally excited compressible gas. Linear equations for inviscid disturbances derived by linearization of the original system (1.27) with respect to a spatially homogeneous steady flow are formulated in Sect. 2.1.

In Sect. 2.2, it is proved by using energy integrals that vibrational relaxation is an additional dissipation factor, which enhances flow stability. Generalization of the Rayleigh's classical first and second theorems is obtained as necessary conditions for instability enhancement in the flows considered. Under certain conditions, a range of eigenvalues of unstable perturbations is specified in the upper complex half-plane as a counterpart of Howard's semicircle theorem. In the limit there is a continuous transition to well-known results for an ideal fluid as the Mach number and the vibrational relaxation time τ approach zero and for an ideal compressible gas as τ approaches zero.

The results of numerical calculations of eigenvalues and eigenfunctions of the most unstable inviscid modes in a free shear layer are presented in Sect. 2.3. Their

Y.N. Grigoryev and I.V. Ershov, *Stability and Suppression of Turbulence in Relaxing Molecular Gas Flows*, Fluid Mechanics and Its Applications 117,
DOI 10.1007/978-3-319-55360-3_2

dependencies on the Mach number of the carrier flow, τ, and the degree of thermal nonequilibrium are analyzed. The vorticity eigenfunctions of these modes are used as initial data for numerical calculations of nonlinear evolution of the Kelvin–Helmholtz waves presented in Chap. 7.

2.1 Equations of the Linear Stability Theory

In the $(x,\ y)$ coordinate plane we consider a shear flow in which the main (carrier) flow of uniform density, ρ_0, and temperature, T_0, is directed along the abscissa axis x and has a velocity profile $U_s = U_s(y)$. The perturbed flow is described by the system of equations of two-temperature gas dynamics (1.27) (see also [2–4]). In dimensionless variables the system has the form

$$\frac{\partial \rho}{\partial t} + \frac{\partial \rho u_i}{\partial x_i} = 0, \quad \rho\left(\frac{\partial u_i}{\partial t} + u_j \frac{\partial u_i}{\partial x_j}\right) = -\frac{\partial p}{\partial x_i},$$

$$\frac{\partial T}{\partial t} + u_i \frac{\partial T}{\partial x_i} + (\gamma - 1) T \frac{\partial u_i}{\partial x_i} + \gamma_v \frac{(T - T_v)}{\tau} = 0,$$

$$\frac{\partial T_v}{\partial t} + u_i \frac{\partial T_v}{\partial x_i} = \frac{T - T_v}{\tau}, \quad \gamma \mathrm{M}^2 p = \rho T, \quad i,\ j = 1,\ 2. \tag{2.1}$$

Here, $x_1 = x$ and $x_2 = y$ and summation over repeated subscripts is implied, ρ is the density, u_i are the velocity components, p is the pressure, and T and T_v are the static and vibrational temperatures of the gas. The parameters in the equations of system (2.1) are defined as follows: $\gamma = c_P/c_V$ is the ratio of specific heats, $c_V = c_{V,tt} + c_{V,rt}$ and $c_P = c_V + R$ are the specific heats at constant volume and constant pressure, due to the translational and rotational motions of molecules, respectively, the parameter $\gamma_v = c_{V,v}/c_V$ represents the degree of excitation of the vibrational mode, $c_{V,v}$ is the specific heat at constant volume corresponding to the relaxing vibrational mode, $\mathrm{M} = U_0/\sqrt{\gamma R T_0}$ is the Mach number of the carrier flow, and R is the gas constant. All the specific heats defined here are assumed to be constant. The scaling parameters are chosen to be a certain characteristic length, L, and a characteristic velocity, U_0, density ρ_0, temperature, T_0, a characteristic time, $\tau_0 = \delta_0/U_0$, and pressure, $p_0 = \rho_0 U_0^2$, derived from these parameters.

System (2.1) describes a situation, widespread in aerodynamics, where the characteristic times of microscopic processes are estimated by the following system of inequalities:

$$\tau_{tt} \sim \tau_{rt} \ll \tau_{vv} \ll \tau_{vt} \sim \tau_0. \tag{2.2}$$

In this case the translational and rotational degrees of freedom, with comparable short relaxation times $\tau_{tt} \sim \tau_{rt}$, form a quasi-equilibrium thermostat with a flow temperature T within times of the order of the characteristic flow time τ_0. Correspondingly, in the subsystem of vibrational levels, a quasi-equilibrium distribution with a vibrational temperature T_v is established within the time τ_{vv}. Relaxation of vibrational degree of freedom to the equilibrium state is described by the Landau-Teller equation with the characteristic time $\tau_{vt} \equiv \tau$. Here and in successive chapters this equation is rewritten in terms of vibrational and translation temperatures T_v, T using relation $E_v = c_{V,v} T_v$ with $c_{V,v} = \text{const}$.

The instantaneous values of the perturbed flow fields are represented as the sum of steady values of the carrier flow and small fluctuations, which depend on the time and the coordinates:

$$u_1 = U_s + u', \quad u_2 = v', \quad \rho = 1 + \rho',$$

$$T = 1 + T', \quad T_v = 1 + T_v', \quad p = \frac{1}{\gamma \mathrm{M}^2} + p'. \tag{2.3}$$

By substituting expressions (2.3) into system (2.1) and linearizing it with respect to the steady flow, we obtain a system of equations for small perturbations.

For our further analysis it is more convenient to pass to an equivalent system of equations for perturbations, in which the equations for density and temperature fluctuations are replaced, with the help of the linearized equation of state, by an equation for pressure fluctuations. The converted system, in which the primes in the notation of the dependent variables are omitted, has the form

$$\frac{\partial u}{\partial t} + U_s \frac{\partial u}{\partial x} + v \frac{\partial U_s}{\partial y} = -\frac{\partial p}{\partial x} = 0, \quad \frac{\partial v}{\partial t} + U_s \frac{\partial v}{\partial x} = -\frac{\partial p}{\partial y} = 0,$$

$$\mathrm{M}^2 \left(\frac{\partial p}{\partial t} + U_s \frac{\partial p}{\partial x} \right) + \frac{\partial u}{\partial x} + \frac{\partial v}{\partial y} = -\frac{\gamma_v}{\gamma} \frac{T - T_v}{\tau},$$

$$\frac{\partial T_v}{\partial t} + U_s \frac{\partial T_v}{\partial x} = \frac{T - T_v}{\tau}, \quad \gamma \mathrm{M}^2 p = \rho + T. \tag{2.4}$$

2.2 Some General Necessary Conditions of Instability Growth

We assume that all the perturbations are periodic along the x axis. This enables us to consider the solutions of system (2.4) in a certain strip along the ordinate axis, finite or unbounded. Then, we can impose zero boundary conditions, both on the finite boundaries of the strip along the y axis and in the asymptotic limits for all the unknown functions. Under these assumptions system (2.4) yields the perturbation energy balance equation

$$\frac{dE}{dt} = -\int uv \frac{\partial U_s}{\partial y} \, d\Omega$$

$$-\frac{\gamma_v}{\gamma^2 \mathrm{M}^2} \int \frac{\rho\,(T - T_v)}{\tau}\, d\Omega - \frac{3}{2}\frac{1}{\mathrm{Re}_r} \int \frac{(T - T_v)^2}{\gamma\,\tau}\, d\Omega, \tag{2.5}$$

where $\mathrm{Re}_r = U_0 L_r / \nu_b$. The integrals here and henceforth are taken over the entire flow region.

The energy integral of the perturbations can be represented in the following quadratic form:

$$E = \frac{1}{2} \int \left(u^2 + v^2 + \mathrm{M}^2\, p^2 + \frac{\gamma_v T_v^2}{\gamma^2 \mathrm{M}^2} \right) d\Omega.$$

Compared to the case of an ideal gas [5], the energy integral E, in addition to the term with the pressure p, defining the perturbation of the internal energy of the gas in the local thermodynamic equilibrium, contains a term with T_v associated with the perturbation of vibrational mode.

The last integral on the right-hand side of Eq. (2.5) is positive definite and shows, in explicit form, the dissipative effect of thermodynamic relaxation of the vibrational mode. In order to emphasize this, by analogy with the equations of energy balance in viscous media [6], we separate the parameter Re_r in front of the integral, which can be defined as the Reynolds relaxation number. It is calculated from the characteristic relaxation "length" $L_r = U_0 \tau$ and the coefficient of kinematic bulk viscosity $\nu_b = \eta_b / \rho_0$, where the coefficient of dynamic bulk viscosity, related to the relaxation vibrational mode, is defined by the relation [2–4]

$$\eta_b = \frac{2}{3} \frac{c_{V,v}}{c_V}\, p_0 \tau.$$

Hence, the presence of thermal relaxation enhances the stability of a compressible plane-parallel flow compared to the case of an ideal gas in local thermodynamic equilibrium [5]. Obviously, a continuous transition from Eq. (2.5) to the analogous equation for an ideal gas occurs. In fact if there is no energy pumping into the vibrational mode and $\tau < \tau_0$, local thermodynamic equilibrium is reached (at $T_v = T$); then, by taking the limit as $\gamma_v \to 0$ and $T_v \to T$, Eq. (2.5) transforms to the equation [5]

$$\frac{dE'}{dt} = -\int u\, v\, \frac{\partial U_s}{\partial y}\, d\Omega, \quad E' = \frac{1}{2} \int \left(u^2 + v^2 + \mathrm{M}^2\, p^2 \right) d\Omega,$$

and only the integral corresponding to energy exchange between the carrier flow and perturbations (Reynolds stresses) remains on the right-hand side.

Because of the presence of the dissipative process in system (2.4), there is no integral of motion similar to that obtained in [5] for an ideal gas. Introducing the perturbation of the generalized potential vorticity in the form

$$\omega = \frac{\partial v}{\partial x} - \frac{\partial u}{\partial y} + \mathrm{M}^2 \frac{\partial U_s}{\partial y}\, p + \frac{\gamma_v}{\gamma} \frac{\partial U_s}{\partial y}\, T_v,$$

from system (2.4) we obtain the equation

$$\frac{\partial \omega}{\partial t} + \frac{\partial U_s \omega}{\partial x} - v\frac{d^2 U_s}{dy^2} = 0$$

which is identical to the potential vorticity equation [5]. The energy equation for the vorticity ω is also identical to the corresponding equation derived previously [5]:

$$\frac{d}{dt}\int U_s \frac{\omega^2}{2}\left(\frac{d^2 U_s}{dy^2}\right)^{-1} d\Omega = \int v\omega U_s \, d\Omega. \tag{2.6}$$

However, further transformations, i.e., integration by parts of the right-hand side of Eq. (2.6) with allowance for the equations of system (2.4), yield the equation

$$\frac{d}{dt}\int\left[U_s \frac{\omega^2}{2}\left(\frac{d^2 U_s}{dy^2}\right)^{-1} + uU_s\left(\mathrm{M}^2 p + \frac{\gamma_v}{\gamma}\cdot T_v\right)\right] d\Omega$$
$$= \int u\, v\frac{\partial U_s}{\partial y}\, d\Omega - \int U_s T_v \frac{\gamma_v}{\gamma}\frac{\partial p}{\partial x}\, d\Omega. \tag{2.7}$$

Unlike the case considered earlier [5], the resultant equation contains additional terms related to the nonequilibrium of the vibrational mode. Addition of the energy equations (2.5) and (2.7) and rearrangement aimed at obtaining complete squares on the left-hand side of the resultant equation lead to the equation

$$\frac{1}{2}\frac{d}{dt}\int\left\{\left[1-(1+\gamma_v)\,\mathrm{M}^2 U_s^2\right] u^2 + v^2 + \mathrm{M}^2\,(p+U_s u)^2\right.$$
$$\left. + U_s\omega^2\left(\frac{d^2 U_s}{dy^2}\right)^{-1} + \gamma_v \mathrm{M}^2\left(U_s u + \frac{T_v}{\gamma \mathrm{M}^2}\right)^2\right\} d\Omega$$

$$= -\frac{\gamma_v}{\gamma^2 \mathrm{M}^2}\int \rho\,\frac{T-T_v}{\tau}\, d\Omega - \int \frac{\gamma_v U_s T_v}{\gamma}\frac{\partial p}{\partial x}\, d\Omega - \frac{3}{2\mathrm{Re}_r}\int \frac{(T-T_v)^2}{\gamma\,\tau}\, d\Omega. \tag{2.8}$$

Here we eliminated the integral describing the exchange of energy between the carrier flow and perturbations. Nevertheless, the time derivative, separated on the left-hand side of Eq. (2.8), cannot vanish, because its right-hand side contains nonzero terms related to the relaxation process, in particular the sign-definite integral, which inhibits the increase in energy of the perturbations with time. However, if local thermodynamic equilibrium is reached in the flow at $\tau < \tau_0$, equality (2.8) in the limit as $T_v \to T$ and $\gamma_v \to 0$ transforms into an integral of a system of linear dynamics of the perturbations in an ideal gas [5]:

$$\frac{1}{2}\frac{d}{dt}\int\left\{\left(1-\mathrm{M}^2U_s^2\right)u^2+v^2\right.$$

$$\left.+\ \mathrm{M}^2\,(p+U_su)^2+U_s\omega^2\left(\frac{d^2U_s}{dy^2}\right)^{-1}\right\}d\Omega=0.$$

Hence, following the well-known approach [5], we find that the perturbation energy E' in the local thermodynamic equilibrium state does not increase if

$$U_s(d^2U_s/dy^2)^{-1}>0,\quad \mathrm{M}\,|U_s|=<1,$$

i.e., under the conditions formulated above, a compressible subsonic flow is stable to small perturbations.

We further consider the stability of wave perturbations in the form of plane waves

$$\boldsymbol{q}=\widetilde{\boldsymbol{q}}\exp\left[i\alpha(x-ct)\right],$$

$$\boldsymbol{q}=(u,\ v,\ \rho,\ p,\ T,\ T_v),\quad \widetilde{\boldsymbol{q}}=(\widetilde{u}(y),\ \widetilde{v}(y),\ \widetilde{\rho}(y),\ \widetilde{p}(y),\ \widetilde{\theta}(y),\ \widetilde{\theta}_v(y)).\qquad(2.9)$$

Here $\boldsymbol{q}$ is the vector of the dependent variables in system (2.4), $\widetilde{\boldsymbol{q}}$ is the vector of the complex amplitudes of the perturbations, $\alpha>0$ is the real wavenumber, and $c=c_r+ic_i$ is the complex phase velocity. Substituting expression (2.9) into system (2.4), we obtain a system of ordinary differential equations for the perturbation amplitudes, for which the following homogeneous boundary conditions are imposed both on the boundaries of the finite interval

$$y_1\le y\le y_2$$

and as $y\to\pm\infty$:

$$\widetilde{u}=\widetilde{v}=\widetilde{\rho}=\widetilde{\theta}=\widetilde{\theta}_v=\widetilde{p}=0.\qquad(2.10)$$

From here we obtain equations for the amplitudes of the pressure perturbations $\widetilde{p}(y)$ and the transverse components of the velocity $\widetilde{v}(y)$, which have the form

$$\left(W^{-2}\widetilde{p}'\right)'-\xi^2\widetilde{p}=0,\quad \xi=\alpha\sqrt{W^{-2}-m^2\mathrm{M}^2},\qquad(2.11)$$

$$\left[\frac{W\widetilde{v}'-U_s'\widetilde{v}}{1-\mathrm{M}^2W^2}\right]'-\alpha^2\,W\widetilde{v}=0.\qquad(2.12)$$

Here

$$W=U_s-c,\quad m^2=m_r^2+im_i^2,$$

$$m_r^2 = \frac{R(1 + \gamma_v + \alpha\tau c_i) + \Delta^2}{R^2 + \Delta^2}, \quad m_i^2 = -\frac{\gamma_v}{\gamma}\frac{(\gamma - 1)\Delta}{R^2 + \Delta^2},$$

$$R = 1 + (\gamma_v/\gamma) + \alpha\tau\, c_i, \quad \Delta = \alpha\tau(U_s - c_r).$$

The prime denotes the operation of differentiation with respect to the y variable.

If there is no excited vibrational mode Eqs. (2.11) and (2.12) are converted by taking the limit as $\gamma_v \to 0$ into equations for the amplitudes of the pressure and transverse velocity perturbations [5]. Although the parameter m^2 is complex, nevertheless, $|m| > 1$, and from the form in which it occurs in Eq. (2.11), we can assume that the action of the relaxation process is similar to the stabilizing action of compressibility. As is well known [5], compressibility reduces the growth rates of the perturbations as compared to the case of an incompressible fluid [7].

An investigation of Eq. (2.11) enables us to obtain some additional stability characteristics. Obviously, for perturbations of the form (2.9) to be unstable, it is necessary that $c_i > 0$. Here $W \neq 0$ and, for the unstable solutions of Eq. (2.11), as in the well-known approach [8], we can introduce a new variable $\widetilde{p} = W^n\, H$. The converted equation in self-conjugate form is

$$\left[W^{2(n-1)}\, H'\right]' + \left[n\, W^n\, \left(W^{n-3} W'\right)' - \xi^2 W^{2n}\right] H = 0. \tag{2.13}$$

The quadratic form of this equation is obtained by multiplying it by the complex conjugate function $\overline{H}$ and integrating over the range of variation of y, taking the boundary conditions (2.11) into account:

$$\int \left\{W^{2(n-1)} |H'|^2 - \left[n W^n \left(W^{n-3} W'\right)' - \xi^2 W^{2n}\right] |H|^2\right\} dy = 0. \tag{2.14}$$

At $n = 0$, it follows from the imaginary part of Eq. (2.14) that

$$2\, c_i \int (U_s - c_r)\left[Q + \alpha^2 \mathrm{M}^2\, Q_1\, |H|^2\right] dy = 0, \tag{2.15}$$

where

$$Q = \frac{|H'|^2 + \alpha^2 |H|^2}{|W|^4}, \quad Q_1 = \frac{\gamma_v(\gamma - 1)\alpha\,\tau}{2\gamma c_i (R^2 + \Delta^2)} > 0.$$

As the expression in the square brackets in the integrand in Eq. (2.15) is non-negative, this equality is satisfied provided that the difference $(U_s - c_r)$ changes its sign in the flow field. Consequently, for developing an instability in the flow of a vibrationally excited gas considered in this study, it is necessary that the Rayleigh condition be satisfied in the same form as for the cases of homogeneous and stratified incompressible fluids and an ideal gas [5]:

$$\min U_s \equiv u < c_r < U \equiv \max U_s. \tag{2.16}$$

Hence, for any enhancement of perturbations the complex phase velocity, c, should lie in the upper half-plane $c_i > 0$ in a half-strip whose width is given by condition (2.16).

However, a more rigid constraint on the phase velocity c, known as the semicircle theorem [1, 5], can only be obtained here under additional conditions.

The real part of Eq. (2.14) has the following form at $n = 0$:

$$\int \left[(U_s - c_r)^2 - c_i^2\right] Q d\, y - K = 0, \quad K = \alpha^2 \mathrm{M}^2 \int m_r^2 |H|^2 d\, y. \tag{2.17}$$

In order to obtain the necessary lower estimate, we consider [1] the inequality

$$0 \geq \int (U_s - u)(U_s - U) Q\, d\, y = I_2 - (U + u) I_1 + U u I_0,$$

$$I_2 = \int U_s^2 Q\, d\, y, \quad I_1 = \int U_s Q\, d\, y, \quad I_0 = \int Q\, d\, y\,. \tag{2.18}$$

The integrals on the right-hand side of the inequality are expressed from Eqs. (2.15) and (2.17) in the form

$$I_1 = c_r I_0 - J, \quad J = \int (U_s - c_r)\, Q_1\, d\, y,$$

$$I_2 = 2 c_r I_1 - \left(c_r^2 - c_i^2\right) I_0 + K\,.$$

Substituting these expressions into inequality (2.18), we obtain

$$0 \geq \left\{ (c_r - \overline{U}\,)^2 + c_i^2 - \frac{(U - u)^2}{4} \right\} I_0 + K + 2(\overline{U} - c_r) J, \quad \overline{U} = \frac{U + u}{2}\,. \tag{2.19}$$

Here $\overline{U}$ is the mean velocity of the carrier flow. If the inequality

$$K + 2(\overline{U} - c_r) J \geq 0, \tag{2.20}$$

holds, we have

$$(c_r - \overline{U}\,)^2 + c_i^2 \leq \frac{(U - u)^2}{4},$$

and the semicircle theorem is also valid [1]: "For any unstable mode at $c_i > 0$, the complex phase velocity lies in the upper half-plane in a semicircle of radius $r = |U - u|/2$ with the center at the point $c_r = \overline{U}$."

In an ideal gas $m_r^2 = 1$ and $Q_1 = 0$; hence, inequality (2.20) is obviously satisfied [5]. Because of the above-mentioned continuity of the transition to the case of an ideal gas, inequality (2.20) and the semicircle theorem also hold for low but finite levels of excitation, as long as the non-sign-definite term in inequality (2.20) does not change the sign of the entire expression. For an arbitrary level of excitation satisfaction of inequality (2.20) must be verified for specific values of the parameters occurring in it.

At $n = 1/2$, the imaginary part of Eq. (2.14) has the form

$$\int |W|^{-2}|H'|^2 dy + \int \left(|W|^{-2} + \alpha^2 m_r^2 \mathrm{M}^2\right)|H|^2 dy + \int \Psi(y)|H|^2 dy = 0,$$

$$\Psi(y) = \frac{1}{c_i}\alpha^2 |m_i^2| \mathrm{M}^2 \left(U_s - c_r\right)$$

$$+\frac{5}{4}\left[3(U_s - c_r)^2 - c_i^2\right]|W|^{-6}U_s'^2 - (U_s - c_r)\,|W|^{-4}]U''.$$

Hence, it follows that, at $c_i > 0$, in order for instability to develop, the function $\Psi(y)$ should change its sign in the interval of integration. Correspondingly, the sufficient condition for stability is expressed (see also [8]) by the requirement that it should be nonnegative.

At $n = 1$ it follows from the imaginary part of Eq. (2.14)

$$c_i \int \Big\{ U_s'' - \left[4\,|W|^{-2}U_s'\right.$$

$$\left. + \alpha^2 \mathrm{M}^2 \left(m_i^2 |W|^4 c_i + 2m_r^2 |W|^2\right)\right](U_s - c_r)\Big\} |H|^2 |W|^{-2}\, dy = 0\,,$$

that the necessary condition for instability is a change in the sign of the expression in the square brackets, within the limits of integration. This requirement extends the well-known Rayleigh condition [1] of the necessity for a point of inflection to exist on the unstable velocity profile in an ideal fluid to the case of a compressible vibrationally excited gas.

2.3 Growth Rates and Eigenfunctions of Unstable Inviscid Modes in a Free Shear Flow

2.3.1 Formulation of the Problem

It is of interest to find the most unstable disturbances of the form (2.15), for which $c_r = U_s(0) = 0$, in a free shear flow with a velocity profile

$$U_s = \tanh\, y, \quad y \subset (-\infty;\ \infty).$$

An eigenvalue problem then arises for Eq. (2.13) with the conditions

$$\widetilde{p}\Big|_{y=\pm\infty} = 0. \tag{2.21}$$

The eigenvalues are c_i, whereas the parameters of the problem are the wavenumber α, the Mach number M, the coefficient γ_v, representing the degree of perturbation of the vibrational mode, and the relaxation time τ. The wavenumber was varied in the range $0 \leq \alpha \leq 1$ in steps of $\Delta\alpha = 0.1$. The Mach number was varied in the subsonic range. The coefficient γ_v took the values $\gamma_v = 0, 0.111, 0.250$, and 0.667. The relaxation time covered a range of three orders of magnitude: $\tau = 0.01, 0.1$, and 1. The calculations were confined to the case of diatomic gases with $\gamma = 1.4$.

We use the methodology of numerical calculations developed earlier when investigating similar eigenvalue problems for an ideal incompressible fluid [7] and an ideal gas [5]. For this velocity profile we have

$$U_s\Big|_{y\to\pm\infty} = \pm 1, \quad U_s{}'\Big|_{y\to\pm\infty} = 0,$$

and for the asymptotic behavior of Eq. (2.13) as $y \to \pm\infty$ we obtain

$$\widetilde{p}'' - \zeta_{\pm}^2 \widetilde{p} = 0,$$

$$\zeta_{\pm} = \alpha\sqrt{1 - m_{\pm}^2 \mathrm{M}^2 (1 \mp c_i)^2}, \quad m_{\pm}^2 = \frac{1 + \gamma_v \pm i\alpha\tau(1 \mp c_i)}{1 + \gamma_v/\gamma \pm i\alpha\tau(1 \mp c_i)}.$$

Hence, the asymptotic solution is defined as

$$\widetilde{p} \sim \exp(\mp\zeta_{\pm} y). \tag{2.22}$$

Using the well-known approach [7], we introduce new dependent and independent variables in problem (2.11) and (2.21):

$$\Pi(y) = d\ln\widetilde{p}/dy = \Pi_r + i\Pi_i, \quad z = \operatorname{th} y.$$

By replacing the variables we transit to a finite interval of integration $z \in [-1;\ 1]$ and reduce the eigenvalue problem to the form

$$\frac{d\Pi}{dz} = \frac{\zeta^2 - \Pi^2}{1 - z^2} + \frac{2\Pi}{z - c_i}, \tag{2.23}$$

$$\Pi\Big|_{z=\pm 1} = \mp\zeta_{\pm}. \tag{2.24}$$

The boundary values (2.24) for the function Π are found from the asymptotic Eq. (2.22). For the calculations it is also necessary to know values of the derivatives of the function Π at the ends of the integration interval. They are found directly from Eq. (2.23) as $z \to \pm 1$. Then, in the first term on the right-hand side of this equation, there is a singularity, which is removed by using L'Hopital's rule. As a result, we have

$$\left.\frac{d\Pi}{dz}\right|_{z=\pm 1} = \frac{\alpha^2 (1 \mp c_i)^2 \Big[2 m_\pm^2 \mathrm{M}^2 \mp B_\pm (1 \mp c_i)\Big] - 4\zeta_\pm}{2(1 \mp c_i)(1 \pm \zeta_\pm)},$$

$$B_\pm = \frac{\gamma_v}{\gamma} \frac{i\alpha\tau(\gamma - 1)\mathrm{M}^2}{\left[1 + \gamma_v/\gamma \pm i\alpha\tau(1 \mp c_i)\right]^2}. \quad (2.25)$$

2.3.2 Numerical Method and Results

To calculate the eigenvalues of the unstable modes Eq. (2.23) and boundary conditions (2.24) and (2.25) are replaced by equations and boundary conditions for the real part Π_r and the imaginary part Π_i of the function Π. The system obtained in this way for fixed sets of parameters is solved numerically by the shooting method. Integration with the Cauchy data (2.24) and (2.25) is carried out by using the fourth-order Runge–Kutta procedure in the ranges $z \in [-1;\ 0]$ and $z \in [0;\ 1]$ in steps of $\Delta z = 10^{-3}$. The "aiming" point is $z = 0$. The values of c are chosen so that the values of the functions Π_r and Π_i calculated "on the left" and "on the right" of the point $z = 0$ are identical to within 10^{-8}. The value of c_i corresponding to this identical value is taken as the eigenvalue for a specified set of parameters. To check the adequacy of the numerical algorithm, we calculate the eigenvalues of the unstable modes in an ideal gas for $\gamma_v = 0$ and $\alpha^2 + \mathrm{M}^2 \leq 1$. The results agree with the data obtained earlier [5] within the doubled accuracy of the computer calculations.

Figure 2.1 shows the isolines of the growth rates αc_i in the $(\mathrm{M},\ \alpha)$ plane for $\tau = 1$ and $\gamma_v = 0.667$ (solid curves). The dashed curves represent the data [5] for an ideal gas. Curves *1* and *2* show the change in the maximum growth rates for an ideal gas and a vibrationally excited gas as a function of the Mach number M. Some numerical values for these curves are given in Table 2.1. The greatest growth rate $\alpha c_i = 0.1897$ is obtained for an ideal fluid with $\mathrm{M} = 0$ and $\alpha = 0.4446$ [7]; for $\mathrm{M} = 1$ the growth rates αc_i are equal to zero.

It can be seen that relaxation, similar to compressibility, reduces the growth rates of the unstable modes. As the Mach number increases, the decrease in the growth rates due to relaxation becomes more and more noticeable. The calculations show that the relaxation effect is enhanced as the excitation level increases, which is specified here by the coefficient γ_v. It should be noted that we used moderate values of γ_v in the calculations. At the same time the model of two-temperature gas dynamics is applicable for higher levels of excitation [2, 3]. We can assume that the relaxation effect on excitation levels close to the onset of dissociation will be comparable in order of magnitude to the decrease in the growth rates with the compressibility effect.

Figure 2.2 shows graphs of the growth rates of the most unstable modes against the relaxation time parameter τ for $\mathrm{M} = 0.5$. In the range of variation of τ and γ_v

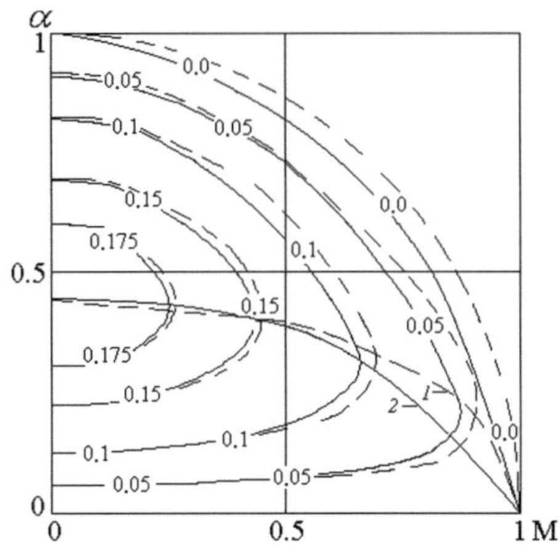

Fig. 2.1 Isolines of the growth rates αc_i at $\tau = 1$. The *dashed* and *solid curves* show the results for $\gamma_v = 0$ and 0.667, respectively. Curves *1* and *2* show the growth rates for $\gamma_v = 0$ and 0.667, respectively

Table 2.1 Spectral characteristics and growth rates of the most unstable inviscid modes for $\tau = 1$, $\gamma_v = 0$ and 0.667

M	α		c_i		$\alpha\, c_i$	
	$\gamma_v = 0$	0.667	0	0.667	0	0.667
0	0.4446	0.4446	0.4266	0.4266	0.1897	0.1897
0.2	0.4260	0.4377	0.4255	0.4115	0.1813	0.1801
0.5	0.3970	0.3890	0.3556	0.3449	0.1413	0.1341
0.8	0.2790	0.2895	0.2790	0.2142	0.0778	0.0620
1	0	0	0	0	0	0

considered in this study the growth rates are almost independent of the relaxation time.

We calculate the eigenfunctions for certain sets of parameters and the eigenvalues of the most unstable modes obtained. We use problem (2.13) and (2.21) to find the eigenfunctions of the pressure. The equation for the eigenfunction of the pressure $\widetilde{p}$ is split into a real and imaginary parts, which are reduced to a first-order system of equations. The asymptotic conditions at $y = \pm\infty$ are transferred to the ends of the interval $[-y_0;\ y_0]$, and the y_0 coordinate is determined from the condition

$$\left|\tanh\, y_0 - 1\right| \le 10^{-12}.$$

The value $y_0 = 20$ is used in the calculations. The system is integrated by using the fourth-order Runge–Kutta procedure.

As the eigenfunctions are determined, except for a constant factor, the normalization of the eigenfunctions of the pressure $\widetilde{p}$ is chosen in the same way as that used

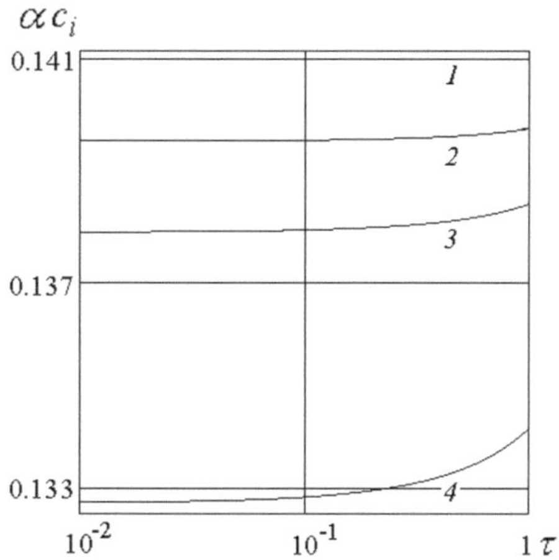

Fig. 2.2 Dependencies of the growth rates αc_i of the most unstable modes against the relaxation time parameter τ for $M = 0.5$ and $\gamma_v = 0$ (*1*), 0.111 (*2*), 0.250 (*3*), and 0.667 (*4*)

previously [5], which enables direct comparisons of the calculated results. Other eigenfunctions are found in terms of the calculated eigenfunctions of the pressures $\widetilde{p}$ and $\widetilde{p}\,'$ by using the relations

$$\widetilde{u} = \frac{U_s{}'\widetilde{p}\,'}{D^2} - \frac{i\alpha\widetilde{p}}{D}, \quad \widetilde{v} = -\frac{\widetilde{p}\,'}{D}, \quad \widetilde{\rho} = \frac{1+\gamma_v+\tau D}{1+\gamma_v/\gamma+\tau D}\,\widetilde{p}\,\mathrm{M}^2,$$

$$\widetilde{\theta} = \left(1+\tau D\right)\widetilde{\theta}_v, \quad \widetilde{\theta}_v = \frac{\gamma-1}{1+\gamma_v/\gamma+\tau D}, \qquad D = i\alpha\,(U_s - c_i) \tag{2.26}$$

following from the system of equations for the perturbation amplitudes.

The curves in Fig. 2.3 illustrate the behavior of the real and imaginary parts of the eigenfunctions of the perturbations of the vibrational temperature $\widetilde{\theta}_v$ of the most unstable modes for $M = 0.5$ and $\tau = 1$. The eigenfunctions of the pressure $\widetilde{p}$ and the eigenfunctions $\widetilde{\rho}$ and $\widetilde{\theta}$, which depend linearly on it, behave in the same way. Their real parts are symmetrical about $y = 0$, while the imaginary parts are antisymmetrical about $y = 0$. The eigenfunctions of the velocity components $\widetilde{u}$ and $\widetilde{v}$, which depend on $\widetilde{p}\,'$, conversely, have antisymmetrical real parts and symmetrical imaginary parts, due to the evenness of the function Π_r and the oddness of the function Π_i. This property, in turn, can be established directly from the system of differential equations for these functions, as was done previously [5, 7].

It is of interest to analyze the effect of excitation on the generalized vorticity perturbation amplitude

$$Real(\widetilde{\omega}) = Real(\widetilde{\omega}_0) + Real(\widetilde{\omega}_1) + Real(\widetilde{\omega}_2)\,.$$

Here

$$Real(\widetilde{\omega}_0) = -(\alpha\,\widetilde{v}_i + \widetilde{u}_r') \cos\alpha\,x - (\alpha\,\widetilde{v}_r - \widetilde{u}_i') \sin\alpha\,x,$$

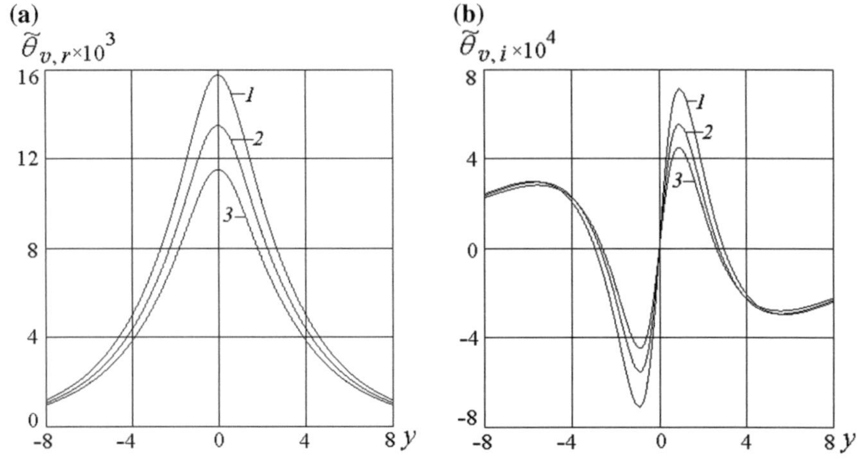

Fig. 2.3 Dependencies of the real $\widetilde{\theta}_{v,r}(y)$ and imaginary $\widetilde{\theta}_{v,i}(y)$ parts of the perturbation of the vibrational temperature $\widetilde{\theta}_v$ for M = 0.5, $\tau = 1$, and $\gamma_v = 0$ (*1*), 0.250 (*2*), and 0.667 (*3*). **a** shows dependencies $\widetilde{\theta}_{v,r}(y)$. **b** shows dependencies $\widetilde{\theta}_{v,i}(y)$

Table 2.2 Numerical absolute values of the real parts of the generalized vorticity $\widetilde{\omega}$ and additive contributions that determine it at $\tau = 1$, $\gamma_v = 0$ and 0.667

M	$\lvert Real(\widetilde{\omega}_0)\rvert$		$\lvert Real(\widetilde{\omega}_1)\rvert$		$\lvert Real(\widetilde{\omega}_2)\rvert$		$\lvert Real(\widetilde{\omega})\rvert$	
	$\gamma_v = 0$	0.667	0	0.667	0	0.667	0	0.667
0.2	0.3992	0.3923	0.0078	0.0082	0	0.0012	0.4072	0.4018
0.5	0.3828	0.3479	0.0412	0.0386	0	0.0056	0.4242	0.3922

$$Real(\widetilde{\omega}_1) = \mathrm{M}^2\, U_S'\, (\widetilde{p}_r \cos\alpha\, x - \widetilde{p}_i \sin\alpha\, x),$$

$$Real(\widetilde{\omega}_2) = \frac{\gamma_v\, U_S'}{\gamma}\, (\widetilde{\theta}_{v,r} \cos\alpha\, x - \widetilde{\theta}_{v,i} \sin\alpha\, x).$$

The corresponding data are presented in Table 2.2.

The isolines of the fluctuations of the generalized (potential) vorticity $\widetilde{\omega}$ in an ideal gas ($\gamma_v = 0$) and in a vibrationally excited gas ($\gamma_v = 0.667$) for M = 0.5 and $\tau = 1$ are compared in Fig. 2.4. Both patterns of the isolines are antisymmetric about the $x = 0$ axis.

According to the data in Table 2.2, excitation and compressibility act in the same direction. Correspondingly, the vorticity decreases as γ_v and M increase, whereas the addition terms due to the internal energy increase. The contribution of excitation of the vibrational mode becomes more significant with increasing Mach number, which was already noted for the growth increments. It is $\widetilde{\omega}_0$ whose absolute value undergoes the greatest change. On the whole, the absolute value of the generalized vorticity decreases as the depth of excitation increases.

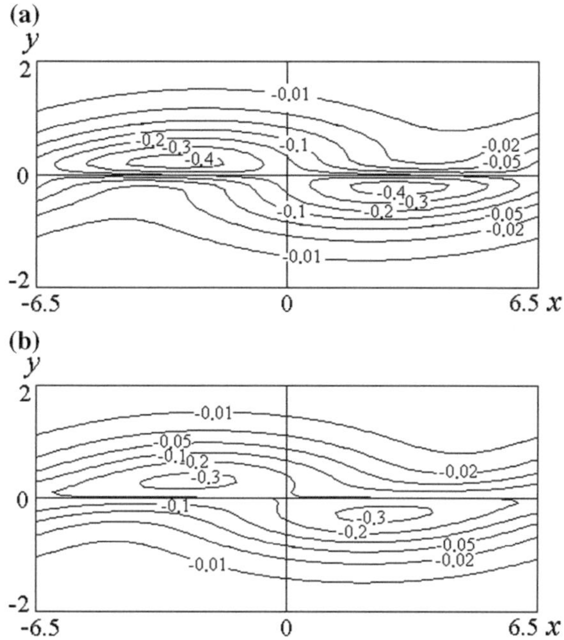

Fig. 2.4 Isolines of fluctuations of the generalized vorticity $\widetilde{\omega}$ for $\mathrm{M} = 0.5$ and $\tau = 1$. **a** is $\gamma_v = 0$. **b** is $\gamma_v = 0.667$

The dissipative effect of excitation can be clearly seen in Fig. 2.4. As compared to the case of an ideal gas, not only the maximum intensity of vortices decreases, but also the gradients of the perturbation field of the generalized vorticity are smoothed.

References

1. Drazin, P.G., Howard, L.N.: Hydrodynamic stability of parallel flow of inviscid fluid. In: Chernyi, G.G., et al. (eds.) Advances in Applied Mechanics, vol. 9, pp. 1–89. Academic Press, New York (1996)
2. Zhdanov, V.M., Aliyevskii, M.Ya.: Transfer and Relaxation Processes in Molecular Gases. Nauka, Moscow (1989) (in Russian)
3. Nagnibeda, E.A., Kustova, E.V.: Non-equilibrium Reacting Gas Flows. Kinetic Theory of Transport and Relaxation Processes. Springer, Berlin (2009)
4. Grigor'ev, Yu.N., Ershov, I.V.: Stability of Flows of Relaxing Molecular Gases. Izd. Sib. Otd. Ross. Akad. Nauk, Novosibirsk (2012) (in Russian)
5. Blumen, W.: Shear layer instability of an inviscid compressible fluid. J. Fluid Mech. **40**, 769–781 (1970)
6. Joseph, D.D.: Stability of Fluid Motion. Springer, Berlin (1976)
7. Michalke, A.: On the inviscid instability of the hyperbolic tangent velocity profile. J. Fluid Mech. **19**, 543–556 (1964)
8. Howard, L.N.: Note on a paper of John W Miles. J. Fluid Mech. **10**, 509–512 (1961)

Chapter 3
Linear Stability of Supersonic Plane Couette Flow of Vibrationally Excited Gas

Abstract The chapter contains the results of numerical and analytical studies of linear stability of a supersonic Couette flow of a vibrationally excited gas. Properties of even and odd inviscid modes of disturbances are analyzed as functions of the Mach number, depth of excitation of vibrational levels, and characteristic relaxation time. The general structure of the spectrum of plane perturbations is studied for finite Reynolds numbers. Two most unstable acoustic viscous modes are identified. Results calculated with using the constant viscosity model and Sutherland's law are compared. Neutral stability curves are obtained, which show that the dissipative effect of vibrational mode excitation is inherent in both models of viscosity. The relative increase in the critical Reynolds number caused by excitation is approximately 12%.

A plane Couette flow is the simplest model of internal flow. Investigations of this flow ensured significant progress in various fields of theoretical hydrodynamics [1]. Stability of the incompressible Couette flow has been the subject of numerous investigations during many years. At the end Romanov [2] provided a rigorous mathematical proof of the absolute linear stability of Couette flow at all Reynolds numbers and arbitrary wavelength of disturbances for incompressible fluid. Such a drastic discrepancy between experimental data and results predicted by the linear stability theory for the laminar-turbulent transition in this flow [3] contributed to the development of the nonlinear energy theory [4] and then of the linear nonmodal stability theory [5].

The situation with the Couette flow of a compressible gas is not so obvious. Application of linear stability theory to study of plane Couette flow with allowance for compressibility has been considered in much fewer papers in compare with incompressible case. Earlier publications (see [6] and references therein) mainly dealt with simplified models; more general results were obtained in [6, 7] and in the recent paper [8], which were united by a common formulation of the problem. The results of asymptotic investigations of stability in the inviscid limit and also at high but finite Reynolds numbers were reported only in [6]. In that case the asymptotic solution for the spectrum of eigenmodes for finite Reynolds numbers was found by using the method of perturbations, which differs from the traditional approach used in the linear theory [9]. In particular the asymptotic construction of the neutral stability curve was not considered.

Y.N. Grigoryev and I.V. Ershov, *Stability and Suppression of Turbulence in Relaxing Molecular Gas Flows*, Fluid Mechanics and Its Applications 117,
DOI 10.1007/978-3-319-55360-3_3

In [6–8] the basic numerical results were obtained by the method of collocations with the QZ-algorithm for finding the spectrum of the phase velocities of disturbances. Nevertheless, the results of [7, 8] contradict the earlier results [6]. Duck et al. [6] revealed a strong stabilizing effect of viscosity and the absence of growing viscous modes up to Reynolds numbers $\mathrm{Re} = 5 \times 10^6$ at Mach numbers $\mathrm{M}_\infty \leq 5$. The absence of growing viscous modes was also found on the basis of asymptotic corrections of the results obtained in the inviscid limit. At the same time the numerical calculations [7, 8] revealed unstable viscous modes at similar Reynolds numbers, though most results were obtained for hypersonic Mach numbers $\mathrm{M}_\infty = 8$–12. Moreover, it was found [7] that viscosity exerts a destabilizing effect in certain ranges of wavelengths, Mach numbers, and Reynolds numbers. In particular a chosen mode that is stable in the inviscid limit becomes unstable.

A possible reason for this difference is the imperfection of implementation of the numerical method in [6], where in-house software was used, in contrast to [7, 8], where professional software was used. In [7] it was additionally tested on the basis of an alternative finite-difference method. However, the inviscid problem was not considered in [7, 8], and there were no asymptotic estimates supporting the results of numerical calculations.

The general characteristics of linear stability of plane-parallel flows of a vibrationally excited gas were considered in Chap. 2 where a significant stabilizing effect of the relaxation process was demonstrated. Linear stability of the Couette flow under conditions of its strong deviation from thermodynamic equilibrium has not been studied until recently. It should be noted that the effect of bulk viscosity characterizing weak excitation of internal degrees of freedom of gas molecules was eliminated in [6–8] by using the known Stokes relation.

These circumstances allowed us to use a supersonic Couette flow for estimating the influence of vibrational excitation on the critical Reynolds number as compared to a nonexcited perfect gas. Simultaneously, we compare the effect of relaxation process for two models of shear viscosity. Because, it was also interested to resolve the contradiction between the results of [6] and [7, 8]. For this purpose, we used the same numerical method in our calculations.

A system of equations for the dynamics of a viscous heat-conducting vibrationally excited gas is formulated in Sect. 3.1. The transport coefficients of this system depend on the static temperature of the flow in the general case. For further analysis, the system is linearized with respect to some solution for the Couette flow in the Navier-Stokes approximation.

Stability of two-dimensional inviscid disturbances in the form of travelling plane waves is considered in Sect. 3.2. For the Couette flow with a linear velocity profile and a parabolic profile of the static temperature, the generalized condition for the inflection point and the sufficient condition in Howard's semicircle theorem, which were obtained in the general case in Chap. 2, are specified here. The complex phase velocities of even and odd inviscid modes of disturbances are numerically calculated as functions of the Mach number, depth of excitation of vibrational levels, and characteristic relaxation time.

The results for stability of disturbances at finite Reynolds numbers are reported in Sect. 3.3. The calculations are carried out for two viscosity models, first, for independent of temperature, and second, for Sutherland's law of temperature dependence.

The dynamics of the overall structure of the spectrum of unstable viscous disturbances is studied. The two most unstable acoustic modes I and II are identified; their inviscid limits were considered in Sect. 3.2. It is shown that viscosity appreciably destabilizes these modes in a certain narrow range of wavenumbers. The influence of viscous stratification of the flow induced by the temperature dependence of viscosity is considered. Neutral stability curves are calculated, which show that the dissipative effect of excitation of the vibrational mode is observed for both models of viscosity. The relative increase in the critical Reynolds number due to excitation is approximately 12% in both considered cases.

3.1 Statement of Problem and Basic Equations

We consider a problem of linear stability of a plane compressible Couette flow of a vibrationally excited molecular gas. In the coordinate plane the flow is bounded by two infinite parallel planes with the distance between them h. It is assumed that the plane $y = 0$ is at rest, and the boundary $y = h$ moves uniformly in its own plane with a velocity U_0. The flow is described within the framework of the model of two-temperature aerodynamics (1.31) (see also [10, 11]). The quantities used for scaling are the channel width h, boundary velocity U_0, density ρ_0 and temperature T_0 of the main flow on the moving boundary of the channel, time $t_0 = L/U_0$, and pressure $p_0 = \rho_0 U_0^2$. In dimensionless variables, the system of equations of two-temperature aerodynamics has the form

$$\frac{\partial \rho}{\partial t} + \frac{\partial \rho u_i}{\partial x_i} = 0, \tag{3.1}$$

$$\rho\left(\frac{\partial u_i}{\partial t} + u_j \frac{\partial u_i}{\partial x_j}\right) = -\frac{\partial p}{\partial x_i} +$$

$$\frac{1}{\mathrm{Re}}\frac{\partial}{\partial x_j}\left[\eta(T)\frac{\partial u_i}{\partial x_j}\right] + \frac{1}{\mathrm{Re}}\left(\alpha_1 + \frac{1}{3}\right)\frac{\partial}{\partial x_i}\left[\eta(T)\frac{\partial u_j}{\partial x_j}\right], \tag{3.2}$$

$$\rho\left(\frac{\partial T}{\partial t} + u_j \frac{\partial T}{\partial x_j}\right) + (\gamma - 1)\rho T \frac{\partial u_i}{\partial x_i} = \frac{\gamma}{\mathrm{Re}\,\mathrm{Pr}}\frac{\partial}{\partial x_i}\left[\eta(T)\frac{\partial T}{\partial x_i}\right] + \frac{\rho\,\gamma_v\,(T_v - T)}{\tau} +$$

$$\frac{\gamma(\gamma-1)\,\mathrm{M}^2\eta(T)}{2\mathrm{Re}}\left[\left(\frac{\partial u_i}{\partial x_j} + \frac{\partial u_j}{\partial x_i}\right)^2 + 2\left(\alpha_1 - \frac{2}{3}\right)\left(\frac{\partial u_i}{\partial x_i}\right)^2\right], \tag{3.3}$$

$$\gamma_v \rho \left(\frac{\partial T_v}{\partial t} + u_j \frac{\partial T_v}{\partial x_j} \right) = \frac{\gamma \alpha_2}{\mathrm{Re\,Pr}} \frac{\partial}{\partial x_i} \left[\eta(T) \frac{\partial T_v}{\partial x_i} \right] - \frac{\gamma_v \rho \, (T_v - T)}{\tau}, \tag{3.4}$$

$$\gamma \mathrm{M}^2 p = \rho T, \quad i = 1,\, 2, \;\; j = 1,\, 2, \tag{3.5}$$

where $x_1 = x$ and $x_2 = y$; summation is performed over repeated indices.

In system (3.1)–(3.5), $\alpha_1 = \eta_b/\eta$ is the ratio of the bulk viscosity to the shear viscosity, $\gamma = c_p/c_V$ is the ratio of specific heats, $c_V = c_{V,t} + c_{V,r}$ and $c_p = c_V + R$ are the specific heats at constant volume and pressure, respectively, which are presented as the sum of specific heats which correspond to translational and rotational motion of molecules, $\gamma_v = c_{V,v}/(c_{V,t} + c_{V,r})$ is the parameter characterizing the degree of nonequilibrium of the vibrational mode, $c_{V,v}$ is the specific heat at constant volume correspond to the relaxing vibrational mode, τ is the characteristic time of relaxation of the vibrational mode, R is the gas constant, $\mathrm{Re} = \rho_0 h U_0/\eta$ and $\mathrm{M} = U_0/\sqrt{\gamma R T_0}$ are the Reynolds and Mach numbers, respectively, $\mathrm{Pr} = \eta c_V/\lambda$ is the Prandtl number, and $\lambda = \lambda_t + \lambda_r$ is the thermal conductivity determined by the translational and rotational degrees of freedom of molecules. Here heat conductivity coefficients are expressed through a shear viscosity coefficient using Eucken's semiempirical relations [12]

$$\lambda_t(T) = \frac{5}{2} c_{V,t}\, \eta(T), \quad \lambda_r(T) = \frac{6}{5} c_{V,r}\, \eta(T), \quad \lambda_v(T) = \frac{6}{5} c_{V,v}\, \eta(T). \tag{3.6}$$

It is assumed the specific heats are independent of static (translational) T and vibrational T_v temperatures and are constant. It is supposed that translational and rotational degrees of freedom are in the quasi-equilibrium state. In this case an equipartition between degrees of freedom takes place and corresponding specific heats are defined by equalities: $c_{V,t} = 3R/2$ and $c_{V,r} = R$. The heat conductivity coefficient $\lambda = \lambda_t + \lambda_r$ used in Eq. (3.3) is defined by summary energy transfer in translational and rotational degrees of freedom.

Parameter α_2 is expressed using Eucken's relations in the form

$$\alpha_2 = \frac{\lambda_{v,0}}{\lambda_{t,0} + \lambda_{r,0}} = \frac{12\gamma_v \left(c_{V,t} + c_{V,r}\right)}{25 c_{V,t} + 12 c_{V,r}} = \frac{20\gamma_v}{33}.$$

In the steady flow an equilibrium state $T_v = \gamma_v = 0$ is assumed, and all parameters depend only on the transverse coordinate y:

$$u_x = U_s(y), \quad u_y = 0, \quad T = T_s(y), \quad \rho = \rho_s(y). \tag{3.7}$$

The following boundary conditions are imposed for the undisturbed flow:

$$U_s(0) = 0, \quad U_s(1) = 1, \quad \left.\frac{dT_s}{dy}\right|_{y=0} = 0, \quad T_s(1) = 1,$$

$$p_s(0) = p_s(1) = 1/\left(\gamma\, \mathrm{M}^2\right). \tag{3.8}$$

Under conditions (3.7), (4.18) the continuity equation (3.1) and the Landau-Teller equation (3.4) are satisfied identically. From the momentum equations (3.2) and equation of state (4.18) it is follows that under boundary conditions (4.18) we have $\rho_s T_s = 1$ and pressure $p_s(y) = 1/(\gamma\, \mathrm{M}^2)$. Simultaneously, from the momentum equations (3.2) and energy equation (3.3) it follows that the velocity profile $U_s(y)$ and temperature profile $T_s(y)$ satisfy the system

$$\frac{d}{dy}\left(\eta_s \frac{dU_s}{dy}\right) = 0,$$

$$\frac{d}{dy}\left(\eta_s \frac{dT_s}{dy}\right) + (\gamma - 1)\, \mathrm{Pr}\, \mathrm{M}^2 \eta_s \left(\frac{dU_s}{dy}\right)^2 = 0, \tag{3.9}$$

where $\eta_s = \eta(T_s)$. Using Satherland's viscosity law [6–8]

$$\eta(T) = \frac{1.5\, T^{3/2}(y)}{T(y) + 0.5}. \tag{3.10}$$

we numerically integrated the system of Eq. (3.9) with boundary conditions (4.18) at $\gamma = 1.4$, Pr = 3/4, by shooting method using the fourth order Runge–Kutta scheme on the interval $y \in [0;\ 1]$.

When shear viscosity is constant one has the exact solution of system (3.1)–(3.5) described by the relations

$$U_s(y) = y, \quad T_s(y) = 1 + \frac{(\gamma - 1)\, \mathrm{Pr}\, \mathrm{M}^2}{2}\left(1 - y^2\right), \quad T_{v,s}(y) = 0,$$

$$\rho_s(y) = \frac{1}{T_s(y)}, \quad p_s(y) = \frac{1}{\gamma \mathrm{M}^2}. \tag{3.11}$$

Figure 3.1 shows the behavior of dependencies of velocity $U_s(y)$ and temperature $T_s(y)$ of mean flow for M = 2 and 5.

To derive linearized equations for small disturbances the instantaneous values of the disturbed flow fields are presented as the sum of the stationary solutions (4.18), (3.9) or (3.11) and small fluctuations, which depend on time and both coordinates in the plane of the flow:

$$u_x = U_s + \widehat{u}_x, \quad u_y = \widehat{u}_y, \quad \rho = \rho_s + \widehat{\rho}, \quad T = T_s + \widehat{T}, \quad T_v = \widehat{T}_v, \quad p = p_s + \widehat{p},$$

$$\eta(T_s + \widehat{T}) = \eta(T_s) + \eta_{T,s}\, \widehat{T} = \eta_s + \widehat{\eta}, \quad \eta_{T,s} = \left.\frac{d\eta}{dT}\right|_{T=T_s}, \tag{3.12}$$

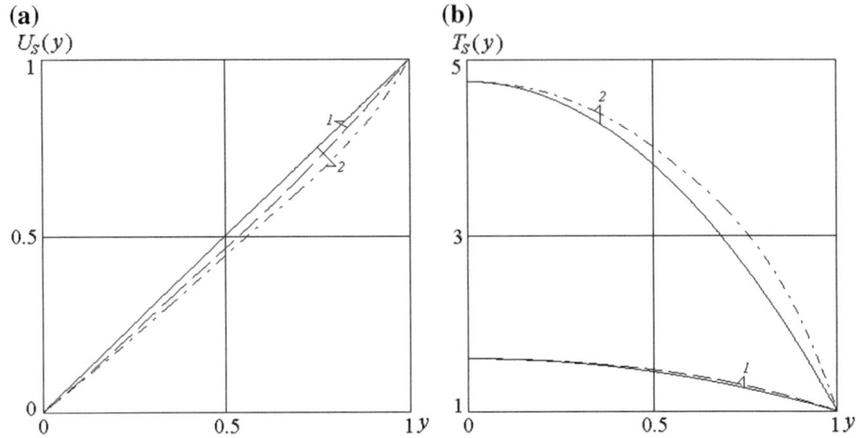

Fig. 3.1 Profiles of velocity $U_s(y)$ and temperature $T_s(y)$ of the mean flow for M = 2 (*1*) and 5 (*2*). **a** shows profiles of velocity $U_s(y)$. **b** shows profiles of temperature $T_s(y)$. The *solid curve* is a constant viscosity model (η = const). The *dashed* and *dashdot curves* are Sutherland's model (3.10)

where additive disturbances have a functional form $\widehat{f} = \widehat{f}(x, y)$.

Substituting Eqs. (3.12) into system (3.1)–(3.5) and linearizing it with respect to the one of the above steady solutions, we obtain the following system of equations for 2D-disturbances:

$$\frac{\partial \widehat{\rho}}{\partial t} + U_s \frac{\partial \widehat{\rho}}{\partial x} + \rho_s \left(\frac{\partial \widehat{u}_x}{\partial x} + \frac{\partial \widehat{u}_y}{\partial y} \right) + \widehat{u}_y \frac{\partial \rho_s}{\partial y} = 0,$$

$$\rho_s \left(\frac{\partial \widehat{u}_x}{\partial t} + U_s \frac{\partial \widehat{u}_x}{\partial x} + \hat{u}_y \frac{\partial U_s}{\partial y} \right) = -\frac{\partial \widehat{p}}{\partial x} + \frac{\eta_s}{\mathrm{Re}} \left(\frac{\partial^2 \widehat{u}_x}{\partial x^2} + \frac{\partial^2 \widehat{u}_x}{\partial y^2} \right)$$

$$+ \frac{\eta_s}{\mathrm{Re}} \left(\alpha_1 + \frac{1}{3} \right) \left(\frac{\partial^2 \widehat{u}_x}{\partial x^2} + \frac{\partial^2 \widehat{u}_y}{\partial x \partial y} \right) + \frac{1}{\mathrm{Re}} \left[\frac{\partial \widehat{u}_x}{\partial y} \frac{d\eta_s}{dy} + \frac{d\widehat{\eta}}{dy} \frac{dU_s}{dy} + \widehat{\eta} \frac{d^2 U_s}{dy^2} \right],$$

$$\rho_s \left(\frac{\partial \widehat{u}_y}{\partial t} + U_s \frac{\partial \widehat{u}_y}{\partial y} \right) = -\frac{\partial \widehat{p}}{\partial y} + \frac{\eta_s}{\mathrm{Re}} \left(\frac{\partial^2 \widehat{u}_y}{\partial x^2} + \frac{\partial^2 \widehat{u}_y}{\partial y^2} \right)$$

$$+ \frac{\eta_s}{\mathrm{Re}} \left(\alpha_1 + \frac{1}{3} \right) \left(\frac{\partial^2 \widehat{u}_x}{\partial x \partial y} + \frac{\partial^2 \widehat{u}_y}{\partial y^2} \right) + \frac{1}{\mathrm{Re}} \left[\left(\alpha_1 + \frac{1}{3} \right) \frac{\partial \widehat{u}_x}{\partial x} + \left(\alpha_1 + \frac{4}{3} \right) \frac{\partial \widehat{u}_y}{\partial y} \right] \frac{d\eta_s}{dy},$$

$$\rho_s \left(\frac{\partial \widehat{T}}{\partial t} + U_s \frac{\partial \widehat{T}}{\partial x} + \widehat{u}_y \frac{\partial T_s}{\partial y} \right) + \gamma(\gamma - 1)\mathrm{M}^2 p_s \left(\frac{\partial \widehat{u}_x}{\partial x} + \frac{\partial \widehat{u}_y}{\partial y} \right)$$

$$= \frac{\gamma \eta_s}{\mathrm{Re\,Pr}} \left(\frac{\partial^2 \widehat{T}}{\partial x^2} + \frac{\partial^2 \widehat{T}}{\partial y^2} \right) + \frac{\gamma_v \rho_s (\widehat{T}_v - \widehat{T})}{\tau} + \frac{2\gamma(\gamma - 1)\mathrm{M}^2 \eta_s}{\mathrm{Re}} \left(\frac{\partial \widehat{u}_x}{\partial y} + \frac{\partial \widehat{u}_y}{\partial x} \right) \frac{dU_s}{dy}$$

$$+\frac{\gamma}{\mathrm{Re\,Pr}}\left[\frac{\partial \widehat{T}}{\partial y}\frac{d\eta_s}{dy}+\frac{\partial \widehat{\eta}}{\partial y}\frac{dT_s}{dy}+\widehat{\eta}\frac{d^2T_s}{dy^2}\right]+\frac{\gamma(\gamma-1)\mathrm{M}^2\widehat{\eta}}{\mathrm{Re}}\left(\frac{dU_s}{dy}\right)^2,$$

$$\gamma_v\rho_s\left(\frac{\partial \widehat{T}_v}{\partial t}+U_s\frac{\partial \widehat{T}_v}{\partial x}\right)=\frac{20\eta_s}{33}\frac{\gamma\gamma_v}{\mathrm{Re\,Pr}}\left(\frac{\partial^2 \widehat{T}_v}{\partial x^2}+\frac{\partial^2 \widehat{T}_v}{\partial y^2}\right)-\frac{\gamma_v\rho_s(\widehat{T}_v-\widehat{T})}{\tau}$$

$$+\frac{20}{33}\frac{\gamma\gamma_v}{\mathrm{Re\,Pr}}\left[\frac{\partial \widehat{T}_v}{\partial y}\frac{d\eta_s}{dy}+\frac{\partial \widehat{\eta}}{\partial y}\frac{dT_s}{dy}+\widehat{\eta}\frac{d^2T_s}{dy^2}\right],$$

$$\gamma\mathrm{M}^2\widehat{p}=\rho_s\widehat{T}+\widehat{\rho}T_s\,. \tag{3.13}$$

Here

$$\frac{d\widehat{\eta}}{dy}=\frac{d\widehat{T}}{dy}\eta_{T,\,s}+\widehat{T}\frac{d\eta_{T,\,s}}{dT}\bigg|_{T=T_s}\frac{dT_s}{dy}.$$

It is assumed that all disturbances vanish on the channel boundaries $y=0$ and $y=1$, and they are periodic with respect to the streamwise coordinate x.

Periodic along x disturbances were considered in the form of travelling plane waves:

$$\boldsymbol{q}(x,\,y,\,t)=\boldsymbol{q}_0(y)\,\exp\left[i\alpha\,(x-ct)\right],$$

$$\boldsymbol{q}(x,\,y,\,t)=(\hat{u}_x,\,\hat{u}_y,\,\hat{\rho},\,\hat{T},\,\hat{T}_v,\,\hat{p}),\ \boldsymbol{q}_0(y)=(u,\,\alpha v,\,\rho,\,\theta,\,\theta_v,\,p)\,. \tag{3.14}$$

Here α is the wavenumber, $c=c_r+ic_i$ is the complex phase velocity, and i is the imaginary unity.

Substitution of expressions (3.14) into system (3.13) gives us a system for amplitudes of disturbances:

$$D\rho+\alpha\rho_s'v+\alpha\rho_s\sigma=0,$$

$$\frac{\eta_s}{\mathrm{Re}}\Delta u-\rho_sDu-\alpha\rho_svU_s'-i\alpha\varepsilon+\frac{\eta_{T,\,s}\,T_s'}{\mathrm{Re}}u'+\frac{\left(\eta_{T,\,s}\,U_s'\right)'}{\mathrm{Re}}\theta+\frac{\eta_{T,\,s}\,U_s'}{\mathrm{Re}}\theta'=0,$$

$$\frac{\alpha\eta_s}{\mathrm{Re}}\Delta v-\alpha\rho_sDv-\varepsilon'+\frac{\alpha\,\eta_{T,\,s}\,T_s'}{\mathrm{Re}}v'=0,$$

$$\frac{\eta_s\gamma}{\mathrm{RePr}}\Delta\theta-\rho_sD\theta-\alpha\rho_svT_s'-\alpha(\gamma-1)\sigma+\frac{2\gamma\,(\gamma-1)\,\eta_s\mathrm{M}^2}{\mathrm{Re}}\left(u'+i\alpha^2v\right)U_s'+$$

$$\frac{\gamma_v\rho_s}{\tau}(\theta_v-\theta)+\frac{2\gamma\,\eta_{T,\,s}\,T_s'}{\mathrm{Re\,Pr}}\theta'+\left[\frac{\gamma(\eta_{T,\,s}\,T_s')'}{\mathrm{Re\,Pr}}+\frac{\gamma(\gamma-1)\,\eta_{T,\,s}\,\mathrm{M}^2\,U_s'^2}{\mathrm{Re}}\right]\theta=0,$$

$$\frac{20}{33}\frac{\gamma\gamma_v\eta_s}{\mathrm{RePr}}\Delta\theta_v-\gamma_v\rho_sD\theta_v-\frac{\gamma_v\rho_s}{\tau}(\theta_v-\theta)+$$

$$\frac{20}{33}\frac{\gamma\gamma_v\eta_{T,s}T_s'}{\mathrm{Re\,Pr}}\theta_v' + \frac{20}{33}\frac{\gamma\gamma_v\eta_{T,s}T_s'}{\mathrm{Re\,Pr}}\theta' + \frac{20}{33}\frac{\gamma\gamma_v\left(\eta_{T,s}T_s'\right)'}{\mathrm{Re\,Pr}}\theta = 0,$$

$$\gamma \mathrm{M}^2 p = \rho_s\theta + \rho T_s,$$

$$D = i\alpha(U_s - c),\quad \sigma = v' + iu,\quad \varepsilon = p - \frac{\alpha\eta_s}{\mathrm{Re}}\left(\alpha_1 + \frac{1}{3}\right)\sigma,\quad \Delta = \frac{d^2}{dy^2} - \alpha^2. \tag{3.15}$$

Here and later the primes mean differentiation with respect to the variable y. Amplitude functions accept zero values on the channel boundaries $y = 0$ and $y = 1$. For a model of a constant shear viscosity system correspondent to (3.15) in a simpler form is presented in Chap. 4.

3.2 Inviscid Stability Problem

In this section we consider stability of the Couette flow of an inviscid vibrationally excited gas. The stationary profiles of velocity and temperature are specified by explicit relations (3.11).

3.2.1 Linear Equations for Inviscid Disturbances

One obtains a system of equations for the amplitude vector $\boldsymbol{q}_0(y)$ of inviscid disturbances by omitting dissipative terms in system (3.15). This is equivalent to the zero approximation in the asymptotic expansion of solution (3.14) with respect to the powers of the small parameter Re^{-1}. The system obtained has the form

$$D\rho + \alpha\rho_s' v + \rho_s\sigma = 0,\quad \rho_s Du + \alpha\rho_s v U_s' + i\alpha p = 0,$$

$$\alpha\rho_s Dv + p' = 0;\quad \rho_s D\theta + \alpha\rho_s v T_s' + (\gamma - 1)\sigma - \frac{\gamma_v\rho_s}{\tau}(\theta_v - \theta) = 0,$$

$$\rho_s D\theta_v + \frac{\rho_s}{\tau}(\theta_v - \theta) = 0,$$

$$\gamma \mathrm{M}^2 p = \rho_s\theta + \rho T_s,\quad D = i\alpha(U_s - c),\quad \sigma = \alpha(v' + iu). \tag{3.16}$$

The amplitude functions take zero values on the channel walls.

3.2.2 Necessary Instability Conditions of Inviscid Modes

After partial elimination of dependent variables, system (3.16) reduces to two equations for the amplitude functions of the disturbances of the transverse velocity and pressure:

$$v' - \frac{vU_s'}{W} - \frac{ip\,T_s}{W}\left[1 - \frac{\mathrm{M}^{*2}\,W^2}{T_s}\right] = 0, \quad p' + \frac{i\alpha^2 W}{T_s}\,v = 0,$$

$$v|_{y=0} = v|_{y=1} = 0, \quad p|_{y=0} = p|_{y=1} = 0. \tag{3.17}$$

Here

$$W = U_s - c, \quad \mathrm{M}^{*2} = m^2\mathrm{M}^2, \quad m^2 = m_r^2 + i\,m_i^2,$$

$$m_r^2 = \frac{R_1\,(1 + \gamma_v + \alpha\tau c_i) + \Delta^2}{R_1^2 + \Delta^2}, \quad m_i^2 = -\frac{\gamma_v}{\gamma}\,\frac{(\gamma - 1)\Delta}{R_1^2 + \Delta^2},$$

$$R_1 = 1 + (\gamma_v/\gamma) + \alpha\tau\,c_i, \quad \Delta = \alpha\tau W_r, \quad W_r = U_s - c_r\,.$$

From Eqs. (3.17), we obtain closed equations for v and p by means of differentiation.

In the self-adjoint form the equation for the amplitude function of the pressure disturbance is written in the following form (cf. [13]):

$$\left(\frac{T_s\,p'}{W^2}\right)' - \alpha^2\left(\frac{T_s}{W^2} - \mathrm{M}^{*2}\right)p = 0. \tag{3.18}$$

Let us consider a quadratic form for Eq. (3.18), which is obtained by multiplying the equation by the complex-conjugate function $\overline{p}$ and integrating with respect to y in the interval [0; 1] with the boundary conditions (3.17). After integration by parts, this form becomes

$$A \equiv \int_0^1 \left[\frac{T_s}{W^2}\,|\,p'|^2 + \alpha^2\left(\frac{T_s}{W^2} - \mathrm{M}^{*2}\right)|\,p\,|^2\right]dy = 0\,. \tag{3.19}$$

The real part of Eq. (3.19) is

$$Real(A) \equiv A_r = \int_0^1 \left[W_r^2 - c_i^2\right]Q\,dy - K = 0,$$

$$K = \alpha^2\mathrm{M}^2 \int_0^1 m_r^2|\,p\,|^2 dy\,, \tag{3.20}$$

and the imaginary part is determined by the expression

$$Imag(A) \equiv A_i = \int_0^1 W_r \left[2c_i Q + \alpha^2 \mathrm{M}^2 Q_1 |\, p\,|^2 \right] dy = 0\,. \tag{3.21}$$

The following notations are used in Eqs. (3.20) and (3.21):

$$Q = \frac{T_s \left(|\, p'|^2 + \alpha^2 |\, p\,|^2 \right)}{|W|^4}, \quad Q_1 = \frac{\gamma_v (\gamma - 1) \alpha\, \tau}{\gamma (R_1^2 + \Delta^2)}\,.$$

For this formulation it is of interest to check satisfaction of the first and second necessary Rayleigh conditions (theorems) of existence of growing inviscid disturbances and also Howard's semicircle theorem [13–16], because these conditions were obtained in [14] and also in [13] for constant profiles of the initial thermodynamic parameters $T_s = \rho_s = 1$, whereas the velocity profile was not specified.

To verify satisfaction of the first Rayleigh condition it is sufficient to note that the expression in the square brackets in Eq. (3.21) is nonnegative. Therefore, for growing disturbances at $c_i > 0$, the imaginary part is $A_i = 0$ if and only if the difference $W_r = U_s - c_r$ changes its sign in the flow field. This means, for the Couette flow with a parabolic temperature profile considered here, the first Rayleigh condition retains the usual form

$$\min U_s \equiv u < c_r < U \equiv \max U_s\,,$$

i.e., for all growing inviscid disturbances the values of the complex phase velocities c should be located in the upper half-plane $c_i > 0$ in the half-band $[u;\ U]$ (in the case considered here, in the interval $[0;\ 1]$).

The second Rayleigh condition (generalized condition of the inflection point [15] (see also [16]) was obtained in Chap. 2 by considering the quadratic form of Eq. (3.19), which is difficult to analyze. As an alternative, one can consider the equation for the disturbance amplitude v, which is presented in the divergent form as

$$\frac{d}{dy}\left[\frac{W v' - v\, U_s'}{\chi} \right] = \frac{\alpha^2 W}{T_s}\, v,$$

$$v|_{y=0} = v|_{y=1} = 0, \quad \chi = T_s - \mathrm{M}^{*\,2} W^2. \tag{3.22}$$

Multiplying Eq. (3.22) by the complex-conjugate function $\overline{v}$ and subtracting the complex-conjugate equation multiplied by v, we derive the equation

$$\overline{v} \frac{d}{dy}\left(\frac{v'}{\chi} \right) - v \frac{d}{dy}\left(\frac{\overline{v}'}{\overline{\chi}} \right) = \frac{|v|^2}{|W|^2} \left[\overline{W} \frac{d}{dy}\left(\frac{U_s'}{\chi} \right) - W \frac{d}{dy}\left(\frac{U_s'}{\overline{\chi}} \right) \right], \tag{3.23}$$

where $\overline{\chi} = T_s - \overline{\mathrm{M}^{*2}}\ \overline{W^2}$.

To simplify further transformations, we should note that, as $c_i \to 0$, one can assume that

$$\chi \approx \widetilde{\chi} = \widetilde{\chi}_r + i\widetilde{\chi}_i = T_s - \widetilde{m}_r^2 \mathrm{M}^2 W_r^2 - i\widetilde{m}_i^2 \mathrm{M}^2 W_r^2$$

$$= T_s - \widetilde{m}_r^2 \mathrm{M}^2 W_r^2 \left(1 + i\frac{\widetilde{m}_i^2}{\widetilde{m}_r^2}\right), \tag{3.24}$$

where

$$\widetilde{m}_r^2 = \frac{\widetilde{R}_1 (1+\gamma_v) + \Delta^2}{\widetilde{R}_1^2 + \Delta^2}, \quad \widetilde{m}_i^2 = -\frac{\gamma_v}{\gamma}\frac{(\gamma-1)\Delta}{\widetilde{R}_1^2 + \Delta^2}, \quad \widetilde{R}_1 = 1 + \frac{\gamma_v}{\gamma}.$$

For the Couette flow of a diatomic gas considered here, we have the estimate

$$\frac{\left|\widetilde{m}_i^2\right|}{\widetilde{m}_r^2} = \frac{\gamma_v}{\gamma}\frac{(\gamma-1)\Delta}{(1+\gamma_v)(1+\gamma_v/\gamma)+\Delta^2} \le \frac{\gamma_v}{\gamma}\frac{\gamma-1}{(1+\gamma_v)(1+\gamma_v/\gamma)} \approx 5\times 10^{-2}.$$

This estimate allows us to neglect the imaginary part of Eq. (3.24) in the first approximation. As a result, Eq. (3.23) is transformed to

$$\frac{d\widetilde{H}}{dy} = \frac{2ic_i|v|^2}{|W|^2}\frac{d}{dy}\left(\frac{U_s'}{\widetilde{\chi}_r}\right), \quad \widetilde{H}(y) = \frac{\overline{v}v' - v\overline{v}'}{\widetilde{\chi}_r}. \tag{3.25}$$

It follows from the last equation that the continuously differentiable function $\widetilde{H}(y)$ takes zero values on the boundaries of the interval of its definition $0 \le y \le 1$. Therefore, according to known theorem of mathematical analysis, its derivative (left side of Eq. (3.25)) should vanish at a certain internal point $\tilde{y}$ of the interval [0; 1]. For growing disturbances $c_i > 0$, it is reached if

$$\frac{d}{dy}\left(\frac{U_s'}{\widetilde{\chi}_r}\right)\bigg|_{y=\tilde{y}} = 0. \tag{3.26}$$

The resultant relation is an obvious generalization of the second Rayleigh condition about the necessity of existence of an inflection point on the unstable profile in an ideal fluid. In [7] the generalized condition of existence of an inflection point is erroneously treated as a necessary condition for neutral stability.

It should be noted that there was an error in [6] in analyzing the equation corresponding to Eq. (3.25) in the case of an ideal gas. For a nonexcited gas we have $\mathrm{M}^* = \mathrm{M}$, and Eq. (3.23) as $c_i \to 0$ takes the form of Eq. (3.25), where

$$\widetilde{\chi}_r = T_s - \mathrm{M}^2 W_r^2.$$

From here, based on similar considerations, there follows a condition of the form (3.26) (cf. (5.11) in [6]). In particular, with given distributions of T_s and U_s (3.11) for an ideal gas, it can be easily shown that the generalized inflection point is determined by the expression

$$\widetilde{y} = c_r \left[1 + \frac{(\gamma - 1)\,\mathrm{Pr}}{2} \right]^{-1}$$

in view of the first Rayleigh condition for c_r, this point is indeed in the interval $0 \le y \le 1$.

In proving the semicircle theorem the initial relation is taken to be the inequality [13, 16]

$$\int_0^1 (U_s - u)(U_s - U) Q \, d\,y = I_2 - (U + u) I_1 + u U I_0 \le 0\,. \tag{3.27}$$

With allowance for a given linear velocity profile the integrals in Eq. (3.27) are determined by the expressions

$$I_0 = \int_0^1 Q\, d\,y, \quad I_1 = \int_0^1 y Q\, d\,y, \quad I_2 = \int_0^1 y^2 Q\, d\,y\,.$$

Which in turn are expressed from Eqs. (3.20) and (3.21) in the form

$$I_1 = c_r I_0 - \frac{J}{2c_i}\,, \quad I_2 = 2c_r I_1 - (c_r^2 - c_i^2) I_0 + K\,,$$

$$J = \alpha^2 \mathrm{M}^2 \int_0^1 (y - c_r) Q_1 |\, p\,|^2 \, d\,y\,.$$

Substituting these expressions into inequality (3.27), one can transform the latter to

$$\left[\left(c_r - \frac{1}{2} \right)^2 + c_i^2 - \frac{1}{4} \right] I_0 + \frac{1}{c_i} \left(\frac{1}{2} - c_r \right) J + K \le 0\,.$$

Obviously, under the condition

$$\left(\frac{1}{2} - c_r \right) J + c_i K \ge 0 \tag{3.28}$$

the values of the complex phase velocity for all unstable inviscid modes at $c_i > 0$ lie in the upper half-plane in a semicircle with a radius $r_0 = 1/2$ and the center at the point $c_r = 1/2$, i.e., condition (3.28) is sufficient for the semicircle theorem to be valid.

In the case of an ideal gas we have $J = 0$ and $c_i K \geq 0$; thus, the semicircle theorem proved in [14] is generalized to the Couette flow with a parabolic temperature profile. It was noted [13] that, by virtue of continuity, inequality (3.28) is also satisfied at small degrees of excitation for $\gamma_v \to 0$; however, inequality (3.28) was included in [13] into the theorem formulation as a sufficient condition in the general case.

The initial velocity and temperature profiles being specified, it is possible to obtain a more correct estimate of condition (3.28). For growing disturbances the first Rayleigh condition in this case is determined by the inequality $0 < c_r < 1$; therefore, condition (3.28) should be considered separately for $0 < c_r < 1/2$ and $1/2 \leq c_r < 1$. At $0 < c_r < 1/2$, condition (3.28) yields the following estimate from below:

$$\left(\frac{1}{2} - c_r\right) J + c_i K = \left(\frac{1}{2} - c_r\right) \alpha^2 \mathrm{M}^2 \int_0^1 \frac{\gamma_v(\gamma - 1)\alpha\,\tau}{\gamma(R_1^2 + \Delta^2)} (y - c_r) |\, p\,|^2 \, d\,y$$

$$+ \alpha^2 \mathrm{M}^2 c_i \int_0^1 \frac{R_1(1 + \gamma_v + \alpha\tau c_i) + \Delta^2}{R_1^2 + \Delta^2} |\, p\,|^2 \, dy$$

$$\geq \left(\frac{1}{2} - c_r\right) \alpha^4 \mathrm{M}^2 \tau^2 c_i \int_0^1 \frac{|\, p\,|^2}{R_1^2 + \Delta^2} \left[\left(y - \frac{1}{2}\right)^2 + \left(y - \frac{1}{2}\right) \frac{\gamma_v(\gamma - 1)}{\gamma\alpha\tau} + \frac{R_1 c_i}{\alpha\tau} \right] dy.$$

It follows from the last integral that inequality (3.28) is satisfied if the quadratic trinomial in the integrand has no real roots in the interval $0 \leq y \leq 1/2$. For this case, the inequality

$$y^2 + ry + q \geq 0, \quad 0 \leq y \leq 1/2,$$

where

$$r = \frac{\gamma_v(\gamma - 1)}{\gamma\alpha\,\tau} - 1, \quad q = \frac{1}{4} - \frac{\gamma_v(\gamma - 1)}{2\gamma\alpha\tau} + \frac{R_1 c_i}{\alpha\tau}.$$

should be satisfied. The condition of complex conjugation of roots (order of roots) is expressed by the inequalities

$$\frac{\gamma_v(\gamma - 1)}{\gamma\alpha\tau} < 1, \quad \frac{R_1}{\alpha\tau} < 1. \tag{3.29}$$

After eliminating the complex $\alpha\tau$ from inequalities (3.29), we obtain the inequality

$$\gamma_v\,(1 - 2/\gamma) < 1,$$

which is always satisfied for the degrees of excitation considered here.

At $1/2 \le c_r < 1$, the estimate from below for inequality (3.28) is obtained in the form

$$\left(\frac{1}{2} - c_r\right) J + c_i K = \left(\frac{1}{2} - c_r\right) \alpha^2 \mathrm{M}^2 \int_0^1 \frac{\gamma_v(\gamma - 1)\alpha\,\tau}{\gamma(R_1^2 + \Delta^2)}(y - c_r)|\,p\,|^2\,d\,y$$

$$+\alpha^2 \mathrm{M}^2 c_i \int_0^1 \frac{R_1(1 + \gamma_v + \alpha\tau c_i) + \Delta^2}{R_1^2 + \Delta^2} |\,p\,|^2\,dy$$

$$\ge \left(c_r - \frac{1}{2}\right) \alpha^4 \mathrm{M}^2 \tau^2 c_i \int_0^1 \frac{|\,p\,|^2}{R_1^2 + \Delta^2}\left[(y - c_r)^2 - (y - c_r)\,\frac{\gamma_v(\gamma - 1)}{\gamma\alpha\tau} + \frac{R_1 c_i}{\alpha\tau}\right] dy.$$

In this case, for inequality (3.28) to be satisfied, it is sufficient that the quadratic trinomial in the integrand does not vanish in the interval $1/2 < y \le 1$. The conditions of complex conjugation of roots (order of roots) of this trinomial also yield inequalities (3.29), which do not impose rigorous conditions on the complex $\alpha\tau$, because inequality (3.28) is satisfied as $\alpha\tau \to 0$. As it follows from inequality (3.29), there is no restriction on the degree of excitation γ_v. It can be added that the proof of the semicircle theorem is actually a matter of prestige in this case, because the calculated values of the imaginary component of the complex velocity (growth rate) for growing modes have the order of 10^{-4}.

3.2.3 *Numerical Collocation Method for Spectral Problem*

System (3.16) with homogeneous boundary conditions is a spectral problem whose eigenvalues are the complex phase velocities of disturbances $c = c_r + ic_i$, and the Mach number M and the wavenumber α serve as parameters. For calculating the eigenvalues $c = c_r + ic_i$ of inviscid modes the spectral problem (3.16) with homogeneous boundary conditions was solved in the Matlab environment. The method of collocations [17, 18] which we previously used in [19, 20] was applied. It should be noted that the method of collocations was also used in [6–8] dealing with investigations of linear stability of the Couette flow in a nonexcited perfect gas. In this case the spectral problem is written in the matrix presentation as follows:

$$A\boldsymbol{\varphi}' + B\boldsymbol{\varphi} = cK\boldsymbol{\varphi}; \quad \boldsymbol{\varphi}\big|_{y=0} = \boldsymbol{\varphi}\big|_{y=1} = 0. \tag{3.30}$$

Here $\boldsymbol{\varphi} = (\rho,\ u,\ v,\ \theta,\ \theta_v)$ is a vector of solution; A, B, and K are 5×5 matrices:

$$A = \begin{pmatrix} 0 & 0 & \rho_s & 0 & 0 \\ 0 & 0 & 0 & 0 & 0 \\ -a_1 & 0 & 0 & -b_1 & 0 \\ 0 & 0 & -c_1 & 0 & 0 \\ 0 & 0 & 0 & 0 & 0 \end{pmatrix},$$

$$B = \begin{pmatrix} -iU_s & -i\rho_s & -\rho_s' & 0 & 0 \\ -i\,a_1 & -i\rho_s U_s & -\rho_s U_s' & -i\,b_1 & 0 \\ -i\,a_1' & 0 & -\alpha\,a_2 & -i\,b_1' & 0 \\ 0 & -ic_1 & -c_2 & -(a_2+b_2) & b_2 \\ 0 & 0 & 0 & b_2 & b_2+\gamma_v a_2 \end{pmatrix},$$

$$K = \begin{pmatrix} -i & 0 & 0 & 0 & 0 \\ 0 & -i\rho_s & 0 & 0 & 0 \\ 0 & 0 & -i\alpha^2\rho_s & 0 & 0 \\ 0 & 0 & 0 & -i\alpha\rho_s & 0 \\ 0 & 0 & 0 & 0 & -i\alpha\gamma_v\rho_s \end{pmatrix},$$

$$a_1 = T_s/\left(\gamma \mathrm{M}^2\right), \quad a_2 = i\alpha\rho_s U_s, \quad b_1 = \rho_s/\left(\gamma \mathrm{M}^2\right),$$

$$b_2 = \gamma_v\rho_s/\tau, \quad c_1 = \alpha\,(\gamma-1), \quad c_2 = \alpha\rho_s T_s'.$$

As collocation nodes, we choose the Gauss–Lobatto points

$$y_n = \frac{1}{2}\left[1+\cos\left(\frac{\pi\,n}{N}\right)\right], \quad n = 0,\ 1,\ \ldots,\ N,$$

in which the Chebyshev polynomial of the Nth power $T_N(y)$ has extremum points in the interval $y \in [0;\ 1]$. The first-order differential operators involved in the spectral problem are approximated on this grid by $(N+1)\times(N+1)$ matrix of collocation derivatives D_N^1 [17, 18]. The elements of this matrix are determined by the formula

$$D^1_{N,lj} = \begin{cases} (-1)^{l+j} s_l/[s_j(z_l - z_j)],\ l \neq j, \\ -z_j/[2(1-z_j^2)],\ 1 \le l = j \le N-1, \\ (2N^2+1)/6,\ l = j = 0, \\ -(2N^2+1)/6,\ l = j = N, \end{cases}$$

$$s_j = \begin{cases} 2,\ j = 0,\ N, \\ 1,\ j = 1,\ 2,\ \ldots,\ N-1. \end{cases} \tag{3.31}$$

This discrete approximation allows problem (3.30) to be reduced to a generalized eigenvalue problem with respect to the spectral parameter c:

$$\sum_{j=0}^{5N+4} (G_{ij} - c\, F_{ij}) z_j = 0, \quad i = 0,\, 1,\, \ldots,\, 5N+4. \tag{3.32}$$

Here, the vector of unknowns z of size $5(N+1)$ consists of the eigenfunction values in the collocation nodes

$$z = (\rho_0,\, \rho_1,\, \ldots,\, \rho_N,\, u_0,\, u_1,\, \ldots,\, u_N,\, v_0,\, v_1,\, \ldots,\, v_N,\\ \theta_0,\, \theta_1,\, \ldots,\, \theta_N,\, \theta_{v,0},\, \theta_{v,1},\, \ldots,\, \theta_{v,N}),$$

the matrices G and F of size $5(N+1) \times 5(N+1)$ are calculated using a special Matlab procedure by the formulas

$$G = A \otimes D_N^1 + B \otimes I_N, \quad F = K \otimes I_N.$$

Here the sign "$\otimes$" means direct (tensor) product of the matrices [21], and I_N is a unit matrix of size $(N+1) \times (N+1)$.

The homogeneous boundary conditions for Eq. (3.32) are taken into account implicitly, via the operator D_N^1 [17, 18]; at the discrete representation, they are obtained by replacing the matrix D_N^1 by a bordered matrix of size $(N-1) \times (N-1)$. The latter is obtained if the following conditions are satisfied:

$$D^1_{0,\,j} = D^1_{N,\,j} = 0, \quad D^1_{i,0} = D^1_{i,\,N} = 0, \quad i = 0,\, \ldots,\, N, \quad j = 0,\, \ldots,\, N.$$

To find all eigenvalues and corresponding eigenfunctions of the generalized spectral problem (3.32), we used the Matlab procedure implementing the QZ-algorithm, which allows the pair of the matrices G and F to be reduced to a generalized upper triangular form by a simultaneous orthogonal transformation. This algorithm is an extension of a better known QR-algorithm, which converts nondegenerate quadratic matrices to the upper triangular form and which is used for solving a standard eigenvalue problem. As a result of applying this procedure for fixed values of the Mach number M, degree of vibrational energy nonequilibrium γ_v, time of vibrational relaxation τ, and wavenumber α, we obtained a set of $N+1$ eigenvalues $c = c_r + ic_i$.

In preliminary calculations the number of collocation nodes on the interval $y \in [0;\ 1]$ was varied from $N+1 = 30$ to $N+1 = 250$. It was observed that variation of the parameter γ_v at $N+1 < 60$ does not affect the imaginary phase velocity c_i which has characteristic value of $c_i \sim 10^{-4}$, due to insufficient precision of computation. Because that in subsequential calculations, the optimal number of collocation nodes on the interval $y \in [0;\ 1]$ was taken to be $N+1 = 152$.

To check the accuracy of the method used, we independently calculated eigenvalues by the "shooting" method. For this purpose Eq. (3.17) were replaced by equations and boundary conditions for the real and imaginary parts of the functions v and p. This system with fixed sets of the parameters M, Pr, γ_v, and τ was numerically integrated by using the fourth-order Runge–Kutta procedure in the intervals $y \in [0;\ 0.5]$ and $y \in [0.5;\ 1]$ with a step $\Delta y = 10^{-3}$. The target point was the middle of the channel $y = 0.5$. The values of c_r and c_i were chosen so that the values of the functions v_r, p_r, v_i, and p_i calculated "on the left" and "on the right" of the point $y = 0.5$ coincided within 10^{-8}. The value of c corresponding to this coincidence was taken as the eigenvalue for a given set of the parameters M, Pr, γ_v, and τ. A comparison of results obtained by the method of collocations and by the shooting method shows that the difference in the values of $c = c_r + ic_i$ is observed only in the sixth-seventh decimal digit. Thus, the necessary accuracy of calculating the disturbance growth rates (decrements) was ensured.

3.2.4 Effect of Vibrational Relaxation on Growth of Second Acoustic Mode

Parametric calculations of the spectral problem showed that variation of the vibrational relaxation times in the interval $10^{-2} \leq \tau \leq 10$ exerts a minor effect on the behavior of the curves $c_r(\alpha, \gamma_v, \mathrm{M})$ and $c_i(\alpha, \gamma_v, \mathrm{M})$. Therefore, the calculated data reported below are given for one characteristic time $\tau = 1$.

The classification of inviscid modes into even and odd modes and the character of their behavior in the case of an ideal gas [6–8] are the same for a vibrationally excited gas [11]. Typical plots of the dependencies $c_r(\alpha)$ for modes of both families are shown in Fig. 3.2. For odd modes, as $\alpha \to 0$ then $c_r > 1$ and $c_r \to \infty$ for all modes except for mode I. At the same time for even modes as $\alpha \to 0$ then $c_r < 0$ and $c_r \to -\infty$ for all modes except for mode II. Modes I and II have finite limits at $\alpha = 0$. For these limits it is possible to obtain an algebraic equation, which allows the values of these limits to be obtained without solving the spectral problem. Indeed, at

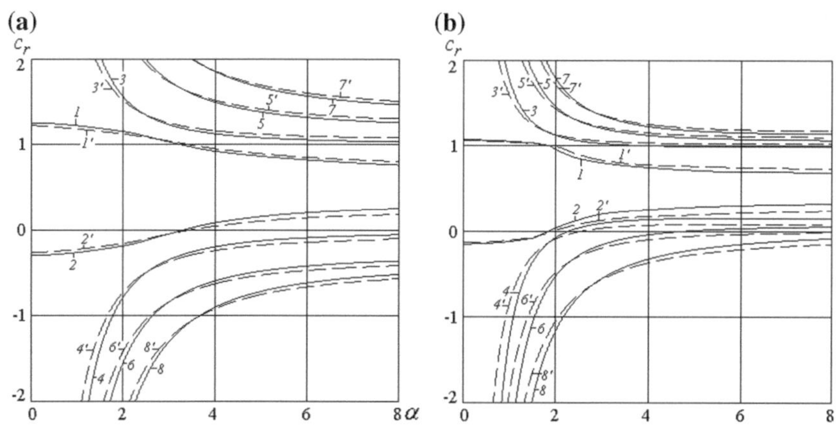

Fig. 3.2 Dependencies $c_r(\alpha)$ for various of the Mach numbers M and $\tau = 1$. **a** is M = 2. **b** is M = 5. Mode I (*1* and *1'*), II (*2* and *2'*), III (*3* and *3'*), IV (*4* and *4'*), V (*5* and *5'*), VI (*6* and *6'*), VII (*7* and *7'*), and VIII (*8* and *8'*). The *solid* and *dashed curves* show the data for $\gamma_v = 0$ and 0.667, respectively

$\alpha = 0$, Eq. (3.22) transforms to a homogeneous differential equation of the second order

$$\frac{d}{dy}\left[\frac{Wv' - vU_s'}{\chi}\right] = 0,$$

which has an obvious first integral in the form of a first-order equation, which is easily integrated. As a result, we obtain the expression for the amplitude of transverse velocity fluctuations

$$v = C\,W \int_0^y \left[\frac{T_s}{W^2} - \mathrm{M}^2 \frac{1+\gamma_v}{1+\gamma_v/\gamma}\right] dy,$$

where the profiles U_s and T_s are defined by Eqs. (3.11) and C is the constant of integration. At $y = 0$ the zero boundary condition (3.17) for v is satisfied. For the boundary condition (3.17) to be satisfied at $y = 1$, following [6], we have to set

$$\int_0^1 \frac{T_s\,dy}{W^2} = \mathrm{M}^2 \frac{1+\gamma_v}{1+\gamma_v/\gamma}. \tag{3.33}$$

For given velocity and temperature profiles (3.11) the integral on the left side of Eq. (3.33) is calculated explicitly. As a result, we receive the nonlinear algebraic equation for the value of c at $\alpha = 0$:

$$\frac{1}{c\,(c-1)} - \frac{(\gamma-1)\mathrm{Pr}\mathrm{M}^2}{2}\left[\frac{1}{c} + 2c\ln\left|\frac{c-1}{c}\right|\right] = \mathrm{M}^2\left[\frac{1+\gamma_v}{1+\gamma_v/\gamma} + (\gamma-1)\mathrm{Pr}\right].$$

The roots of this equation are found by the Newton method. At $\mathrm{M} = 2$ and $\gamma_v = 0$, we have

$$c_r^{\mathrm{I}} = 1.2464, \quad c_i^{\mathrm{I}} = -0.074490 \times 10^{-10}$$

and

$$c_r^{\mathrm{II}} = -0.2935, \quad c_i^{\mathrm{II}} = -0.864249 \times 10^{-10},$$

at $\gamma_v = 0.667$, we obtain correspondingly

$$c_r^{\mathrm{I}} = 1.2211, \quad c_i^{\mathrm{I}} = -0.181863 \times 10^{-10}$$

and

$$c_r^{\mathrm{II}} = -0.2663, \quad c_i^{\mathrm{II}} = -0.685248 \times 10^{-10}.$$

The absolute deviation of the found roots from the corresponding eigenvalues obtained in solving the spectral problem (3.16) does not exceed 10^{-8}. Deviations of the same order were obtained for other values of the parameters. This fact confirms

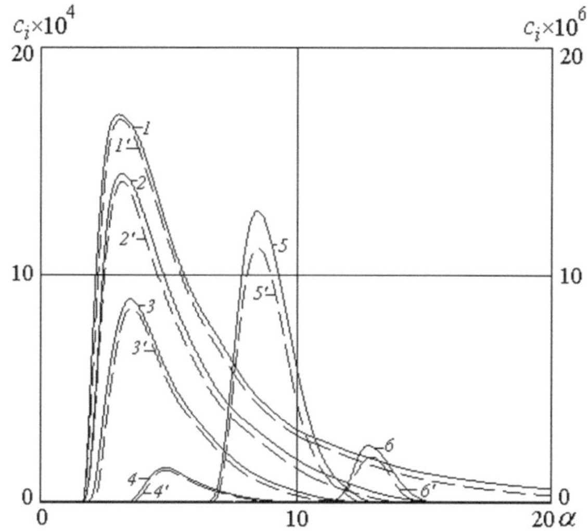

Fig. 3.3 Dependencies $c_i(\alpha)$ for the most unstable mode II at M $= \infty$ (*1* and *1′*), 20 (*2* and *2′*), 10 (*3* and *3′*), 5 (*4* and *4′*), 2 (*5* and *5′*), 0,5 (*6* and *6′*), and 0 (*7* and *7′*). The *solid* and *dashed curves* show the results for $\gamma_v = 0$ and 0.667, $\tau = 1$, respectively

the accuracy of eigenvalue calculations. It follows from Fig. 3.2 that the values of c_r for all modes in the ranges of wavenumbers $0 \leq \alpha < 3.4$ at M $= 2$ and $0 \leq \alpha < 2$ at M $= 5$ are outside the interval [0; 1], in which instability development is possible in accordance with the first Rayleigh condition. Stability of all modes in this range of wavenumbers is confirmed by the corresponding negative calculated values of c_i. At the same time we have $|c_i| \ll |c_r|$ in our case, which makes stability of these modes close to neutral stability. Nevertheless, the calculations show that modes I and II in the interval $0 \leq \alpha < 3.4$ are not neutrally stable, as argued in [6, 7].

At $c_r < 1$ mode II becomes unstable in the entire range of Mach numbers (see Fig. 3.3). In contrast to [6] the calculated curves in the range of wavenumbers $\alpha = 8$–10 have no peaks that do not possess a physical meaning. Even modes with greater numbers also have positive values $c_i > 0$, which, however, are quantities of the second order of smallness with respect to mode II. Odd modes I, III, etc., remain stable as long as the first Rayleigh condition $0 < c_r < 1$ is satisfied for them. At fixed Mach numbers the relations $c_i^{\mathrm{II}} \simeq |c_i^{\mathrm{I}}|$ are valid in the range of wavenumbers α where the maximum values of $c_i > 0$ are reached for mode II.

Calculations of linear stability of a free shear layer with respect to inviscid disturbances in an ideal gas and a vibrationally excited gas presented in Chap. 2 (see also [13, 14]) show that an increase in the Mach number leads to a decrease in the values of the imaginary part of the phase velocity c_i (growth rates $\omega = \alpha c_i$). At the same time it was found for the Couette flow of an ideal gas [6] that the growth rates of the most unstable mode II, vice versa, increase with increasing compressibility and asymptotically tend to a certain finite limit as M $\to \infty$. A similar behavior of the growth rates is observed for a vibrationally excited gas (see Fig. 3.3). The scale of the values of c_i for M $= 0.5$ and M $= 0$ is located on the right in Fig. 3.3.

In the limiting cases, $\mathrm{M} \to \infty$ and $\mathrm{M} = 0$, the spectral problem (3.22) is simplified. At large values of M the expression for the temperature profile is approximately presented as

$$T_s \to \mathrm{M}^2 \widetilde{T}_s(y), \quad \widetilde{T}_s(y) = \frac{(\gamma - 1)\mathrm{Pr}}{2}\left(1 - y^2\right).$$

As $\mathrm{M} \to \infty$ it follows from Eq. (3.22) that

$$\frac{d}{dy}\left[\frac{Wv' - vU_s'}{\widetilde{T}_s - m^2 W^2}\right] = \frac{\alpha^2 W}{\widetilde{T}_s} v, \quad v|_{y=0} = v|_{y=1} = 0. \tag{3.34}$$

At $\mathrm{M} = 0$ Eq. (3.22), in the case of an ideal incompressible fluid, yields a variant of the Rayleigh equation [22] for the amplitude of the transverse velocity disturbance

$$\left(Wv' - vU_s'\right)' - \alpha^2 Wv = 0, \quad v|_{y=0} = v|_{y=1} = 0. \tag{3.35}$$

The limiting growth rates obtained from Eqs. (3.34) and (3.35) are also shown in Fig. 3.2.

It is seen in Fig. 3.3 that vibrational nonequilibrium described by the parameter γ_v leads to decreases in the values of the imaginary parts of the phase velocities $c_i(\alpha, \gamma_v)$ as compared to the values of $c_i(\alpha)$ for an ideal gas ($\gamma_v = 0$). Table 3.1 shows the maximum values of the imaginary parts of the phase velocity $c_i^{\max}$ and the growth rates $\omega^{\max} = \alpha c_i^{\max}$ in a wide range of Mach numbers M at $\gamma_v = 0$ and 0.667 at $\tau = 1$. It should be noted that the data obtained are only qualitative at high Mach

Table 3.1 Maximum growth rates $\omega^{\max}$ and corresponding wavenumbers α

M	Ideal gas ($\gamma_v = 0$)			Vibrationally excited gas ($\gamma_v = 0.667$, $\tau = 1$)		
	α	$c_i^{\max} \times 10^4$	$\omega^{\max} \times 10^3$	α	$c_i^{\max} \times 10^4$	$\omega^{\max} \times 10^3$
0.0	12.9870	0.248	0.0322	12.9760	0.0198	0.0257
0.5	8.5405	0.1296	0.1107	8.5476	0.1102	0.0942
1.0	6.4498	0.4112	0.2652	6.4504	0.3485	0.2248
2.0	4.9681	1.5962	0.7930	4.9713	1.4002	0.6961
3.0	4.3550	4.2418	1.8473	4.3575	3.3785	1.4722
4.0	3.9102	6.7012	2.6203	3.9186	5.8698	2.3001
5.0	3.5973	8.8636	3.1885	3.6017	8.3005	2.9896
10.0	3.1250	14.2480	4.4525	3.1295	13.9480	4.3650
20.0	3.0055	16.2490	4.8836	2.9918	16.0490	4.8016
30.0	2.9690	16.5490	4.9134	2.9710	16.2490	4.8476
40.0	2.9644	16.6980	4.9499	2.9698	16.3980	4.8699
∞	2.9555	17.0000	5.0244	2.9585	16.5000	4.8815

numbers $M \geq 20$ at which the real gas effects are essential and both models have no physical meaning. At the same time the results obtained show that the effect of reduction of growth rates of unstable inviscid modes on the background of the relaxation process is clearly expressed in the Couette flow.

3.3 Linear Stability of Supersonic Couette Flow at Finite Reynolds Numbers

In this section the influence of vibrational relaxation on the evolution of viscous disturbances is investigated. Two viscosity models are simultaneously considered. Correspondingly, stationary profiles of velocity and temperature defined by numerical integration of system (3.9) or by explicit relations (3.11) are used in linearized amplitude equations (3.15).

3.3.1 Numerical Calculations of Spectral Problem

System (3.15) with homogeneous boundary conditions on the channel walls $y = 0$ and $y = 1$ defines a spectral problem whose eigenvalues are the complex phase velocities of the viscous disturbances $c = c_r + ic_i$, while the Reynolds number Re, Mach number M, and wavenumber α serve as parameters. For calculating the eigenvalues of the viscous modes we also used the collocation method defined in Sect. 3.2.3. In the present study the spectral problem is written in the matrix presentation as

$$A_1\boldsymbol{\varphi}'' + A_2\boldsymbol{\varphi}' + A_3\boldsymbol{\varphi} = cA_4\boldsymbol{\varphi}, \quad \boldsymbol{\varphi}\big|_{y=0} = \boldsymbol{\varphi}\big|_{y=1} = 0, \tag{3.36}$$

Here $\boldsymbol{\varphi} = (\rho, u, v, \theta, \theta_v)$ and A_k $(k = 1, 2, 3, 4)$ are 5×5 matrices:

$$A_1 = \begin{pmatrix} 0 & 0 & 0 & 0 & 0 \\ 0 & \dfrac{\eta_s}{\text{Re}} & 0 & 0 & 0 \\ 0 & 0 & \dfrac{\alpha\eta_s}{\text{Re}}\left(\alpha_1 + \dfrac{4}{3}\right) & 0 & 0 \\ 0 & 0 & 0 & \dfrac{\gamma\eta_s}{\text{RePr}} & 0 \\ 0 & 0 & 0 & 0 & \dfrac{20\gamma\gamma_v\eta_s}{33\text{RePr}} \end{pmatrix},$$

$$A_2 = \begin{pmatrix} 0 & 0 & -\rho_s & 0 & 0 \\ 0 & a_1 & \dfrac{i\alpha^2\eta_s}{\text{Re}}\left(\alpha_1 + \dfrac{1}{3}\right) & a_5 & 0 \\ -\dfrac{T_s}{\gamma \text{M}^2} & \dfrac{i\alpha\eta_s}{\text{Re}}\left(\alpha_1 + \dfrac{1}{3}\right) & \alpha\, a_2 & -\dfrac{\rho_s}{\gamma \text{M}^2} & 0 \\ 0 & \dfrac{2\gamma\,(\gamma-1)\,\eta_s \text{M}^2 U_s'}{\text{Re}} & -\alpha\,(\gamma-1) & a_3 & 0 \\ 0 & 0 & 0 & a_4 & a_4 \end{pmatrix},$$

$$A_3 = \begin{pmatrix} -iU_s & -i\rho_s & -\rho_s' & 0 & 0 \\ -\dfrac{i\alpha T_s}{\gamma \text{M}^2} & -\alpha\, b_1 & -\alpha\rho_s U_s' & -\dfrac{i\alpha\rho_s}{\gamma \text{M}^2} + \dfrac{\left(\eta_{T,s}\, U_s'\right)'}{\text{Re}} & 0 \\ -\dfrac{T_s'}{\gamma \text{M}^2} & i\alpha\, b_5 & -\alpha^2\, b_2 & -\dfrac{\rho_s'}{\gamma \text{M}^2} & 0 \\ 0 & -i\alpha\,(\gamma-1) & \alpha\, b_6 & -b_3 & \dfrac{\gamma_v \rho_s}{\tau} \\ 0 & 0 & 0 & \gamma_v\, b_7 & -\gamma_v\, b_4 \end{pmatrix},$$

$$A_4 = \begin{pmatrix} -i & 0 & 0 & 0 & 0 \\ 0 & -i\alpha\rho_s & 0 & 0 & 0 \\ 0 & 0 & -i\alpha^2\rho_s & 0 & 0 \\ 0 & 0 & 0 & -i\alpha\rho_s & 0 \\ 0 & 0 & 0 & 0 & -i\alpha\gamma_v\rho_s \end{pmatrix},$$

$$a_1 = \frac{\eta_{T,s}\, T_s'}{\text{Re}}, \quad a_2 = a_1\left(\alpha_1 + \frac{4}{3}\right), \quad a_3 = \frac{2\gamma\, a_1}{\text{Pr}}, \quad a_4 = \frac{10\gamma_v a_3}{33},$$

$$a_5 = \frac{\eta_{T,s}\, U_s'}{\text{Re}}, \quad b_1 = i\rho_s U_s + \frac{\alpha\eta_s}{\text{Re}}\left(\alpha_1 + \frac{4}{3}\right), \quad b_2 = i\rho_s U_s + \frac{\alpha\eta_s}{\text{Re}},$$

$$b_3 = i\alpha\rho_s U_s + \frac{\gamma_v\rho_s}{\tau} + \frac{\alpha^2\gamma\eta_s}{\text{RePr}} - \frac{\gamma\, a_1'}{\text{Pr}} - \gamma(\gamma-1)\,\text{M}^2\, a_5\, U_s',$$

$$b_4 = i\alpha\rho_s U_s + \frac{\rho_s}{\tau} + \frac{20\gamma\,\eta_s\alpha^2}{33\text{RePr}}, \quad b_5 = a_1\left(\alpha_1 + \frac{1}{3}\right),$$

$$b_6 = \frac{2i\alpha\gamma\,(\gamma-1)\,\text{M}^2\eta_s U_s'}{\text{Re}} - \rho_s T_s', \quad b_7 = \frac{\gamma_v\rho_s}{\tau} + \frac{20\gamma_v\gamma\left(\eta_{T,s}\, T_s'\right)'}{33\text{RePr}}.$$

We checked the accuracy of solution using the shooting method as done in the inviscid spectral problem. Comparison of the results obtained by the collocation and shooting methods shows that the differences in the values of $c = c_r + ic_i$ are observed only in the sixth or seventh decimal digits. Thus, the necessary accuracy of calculating the growth rates of disturbances was provided.

The neutral curves $c_i(\alpha, \mathrm{Re}) = 0$ for the nth viscous mode of disturbances are calculated in the following way. At fixed values of the parameters α_1, γ_v, τ_{vt}, and Mach number M, we calculate two-dimensional arrays of the growth rates of the nth viscous mode

$$\omega^n_{jk} = \omega^n\left(\alpha_j, \mathrm{Re}_k\right) = \alpha_j c_i^n\left(\alpha_j, \mathrm{Re}_k\right), \tag{3.37}$$

where the one-dimensional arrays α_j and Re_k are defined by the formulas:

$$\alpha_j = \alpha_0 + j\,\Delta\alpha, \quad j = 0, 1, \ldots, J,$$

$$\mathrm{Re}_k = \mathrm{Re}_0 + k\,\Delta\mathrm{Re}, \quad k = 0, 1, \ldots, K,$$

$\Delta\alpha$ and $\Delta\mathrm{Re}$ are the increments of the wavenumber and Reynolds number, respectively.

Array (3.37) defines the surface $\omega(\alpha, \mathrm{Re})$ for the nth viscous mode of disturbances. The coordinates in the set defining the geometric place of the points of this level line $\omega(\alpha, \mathrm{Re}) = C$ on the plane (α, Re) are found from the inequality

$$\left|\omega^n\left(\alpha_j, \mathrm{Re}_k\right) - C\right| \leq 10^{-8},$$

where C is a certain prescribed number. For $C = 0$ we obtain the neutral curve $\omega(\alpha, \mathrm{Re}) = 0$ for the nth viscous mode in terms of the growth rate of disturbances.

The computations are performed for the following values of the parameters:

$$\gamma_v = 0\text{–}0.667, \quad \tau = 0.01\text{–}10, \quad \mathrm{M} = 0.5\text{–}25, \quad \mathrm{Re} = 10^4\text{–}10^8,$$

$$\alpha_1 = 0\text{–}2, \quad \alpha = 0\text{–}10, \quad \Delta\mathrm{Re} = 50, \quad \Delta\alpha = 10^{-3}, \quad \mathrm{Pr} = 3/4, \quad \gamma = 7/5.$$

The number of collocation nodes in the, interval $y \in [0; 1]$ is varied from $N + 1 = 100$ to $N + 1 = 500$; in most computations, it was taken to be $N + 1 = 300$. This choice is determined by the sensitivity of the eigenvalues of the viscous problem to computation errors, which was noted in [7, 8].

3.3.2 Structure of Spectra of Viscous Disturbances

The parametric calculations of the considered spectral problem for the constant viscosity model and Satherland's viscosity model show that variations of the vibrational relaxation time in the interval $10^{-2} \leq \tau \leq 10$ weakly affect the character of the spectra of viscous disturbances. Therefore, the data given below are presented for one value of the relaxation time $\tau = 1$.

For the Couette flow of a perfect gas it was shown [6, 7] that the spectrum of the phase velocities of viscous disturbances contains two characteristic sets S_a and S_v, which are generalizations of the splitting obtained in [23] for incompressible flows. Each eigenvalue in S_a converges as $\mathrm{Re} \to \infty$ to a certain eigenvalue for one of the inviscid modes. Eigenvalues from the set S_v have no asymptotic relationship with the inviscid problem and are characterized by large damping decrements. The calculations of viscous disturbances show that this structure of the spectrum is also observed in the case of a vibrationally excited gas. Thus, splitting of the set S_a into even and odd modes, which was obtained in [6, 7] for inviscid disturbances in equilibrium gas and was presented in Sect. 3.2.4 in vibrationally excited gas, is also valid for viscous disturbances. In particular the behavior of modes I and II, which have finite limits $c_r(\alpha)$ as $\alpha \to 0$, is retained. As there are several possible interpretations of the physical meaning of these and other modes of the set S_a, the following comment should be made. In contrast to the known modes of disturbances in a compressible boundary layer, where the first mode is an extension of the Tollmien–Schlichting wave to the case of a compressible flow and the second mode is an acoustic mode, both families of modes consist of acoustic modes [24].

Dependencies of wave velocities $c_r(\alpha)$ for even and odd modes of perturbation from set S_a are presented in Fig. 3.4 for two values of the Mach number $\mathrm{M} = 3$ and 5. Here the solid and dashed lines show dependencies $c_r(\alpha)$ for even and odd inviscid modes of perturbation for ideal ($\gamma_v = 0$) and vibrationally excited ($\gamma_v = 0.667$) gases, respectively. The dots present the results for both viscosity models under two Reynolds numbers $\mathrm{Re} = 10^5$ and 10^6. The values for both models with vibrational excitation are not shown on the curves in order to simplify graphs. Analysis of Fig. 3.4 allows one to conclude dissipative effects related with "viscous" stratification, variation the Reynolds number Re, and vibrational excitation do not change practically a behavior of phase velocity curves $c_r(\alpha)$ calculated in ideal gas approximation under $\mathrm{Re} \to \infty$ (see Fig. 3.3).

The curves $c_r(\alpha)$ in Fig. 3.4 show that modes I and II enter the interval $c_r = [0;\ 1]$, where the development of the inviscid instability is possible, earlier than other modes.In this sense modes I and II are the most unstable ones.

At $\mathrm{M} = 3$ the curves for modes I and II of both models are almost mirror-symmetric with respect to the line $c_r = 0.5$ corresponding to the position of the critical layer on the channel axis. It should be noted that the points of the transition of modes I and II to the interval $c_r = [0;\ 1]$ are shifted with increasing Mach number to the domain of smaller wavenumbers α, and the symmetry with respect to the line $c_r = 0.5$ is violated. This is caused by the fact that the temperature profiles T_s deviate substantially from symmetric form as Mach number increases. Simultaneously, the viscous stratification does not exert additional visible effect on deviation from symmetry despite the greater asymmetry of the temperature profile compared with the constant viscosity model.

The general structure of the spectrum of viscous disturbances for different flow parameters is illustrated in Figs. 3.5 and 3.6. The majority of the spectral points on the complex plane $(c_r,\ c_i)$ belongs to the set S_v of intensely damping, purely viscous disturbances. All of them are located in the lower half-plane within the

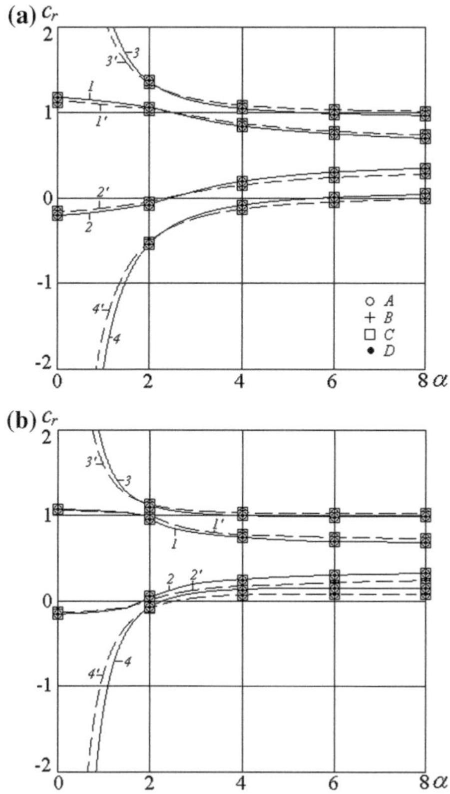

Fig. 3.4 Dependencies $c_r(\alpha)$ for various of the Mach numbers M at $\alpha_1 = 0$. **a** is $\mathrm{M} = 3$. **b** is $\mathrm{M} = 5$. The *solid* and *dashed curves* are inviscid modes at $\gamma_v = 0$ and 0.667, respectively. *A* is a constant viscosity model at $\mathrm{Re} = 10^5$. *B* is a constant viscosity model at $\mathrm{Re} = 10^6$. *C* is Sutherland's model (3.10) at $\mathrm{Re} = 10^5$. *D* is Sutherland's model (3.10) at $\mathrm{Re} = 10^6$. Mode I (*1* and *1'*), II (*2* and *2'*), III (*3* and *3'*), IV (*4* and *4'*)

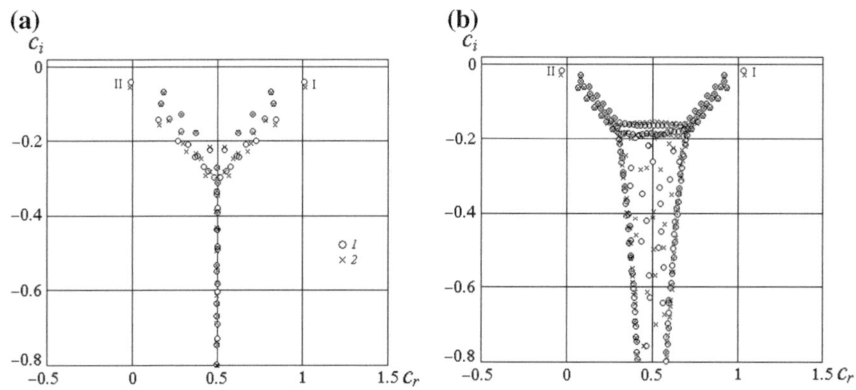

Fig. 3.5 Spectra of the eigenvalues $c = c_r + ic_i$ for $\mathrm{M} = 3$, $\alpha = 0.1$ and $\gamma_v = 0$. **a** is $\mathrm{Re} = 10^5$. **b** is $\mathrm{Re} = 10^6$. $\alpha_1 = 0$ (*1*) and 2 (*2*). Points I and II show the results for modes I and II, respectively

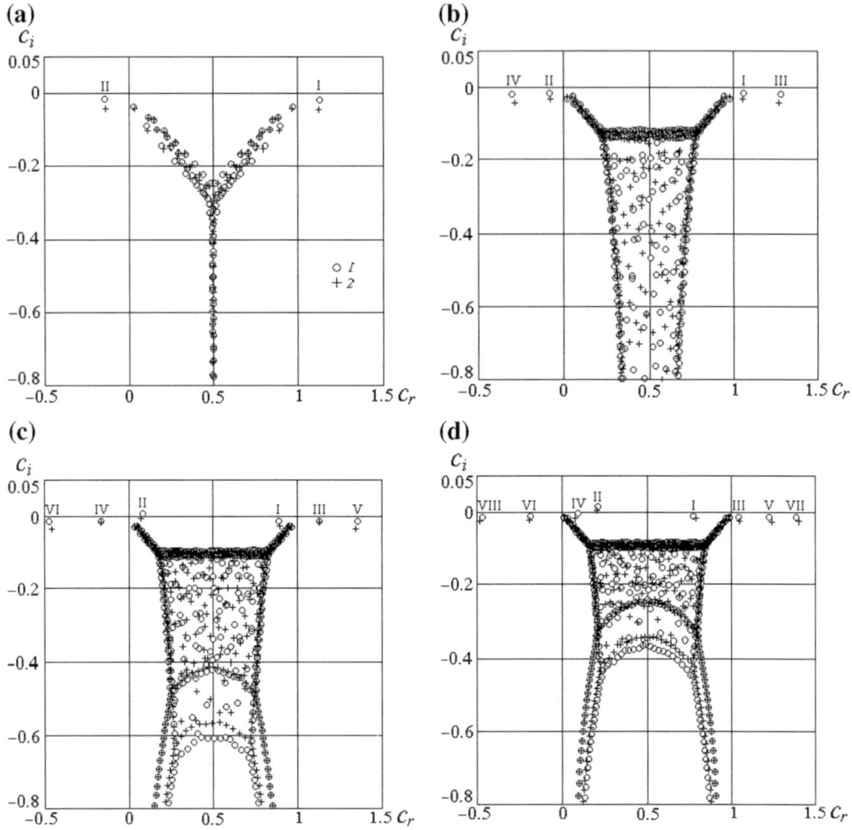

Fig. 3.6 Spectra of the eigenvalues $c = c_r + ic_i$ for Re $= 2.5 \times 10^5$, M $= 5$. **a** is $\alpha = 0.1$. **b** is $\alpha = 1$. **c** is $\alpha = 2$. **d** is $\alpha = 3$. $\alpha_1 = \gamma_v = 0$ (*1*), $\alpha_1 = 2$ and $\gamma_v = 0.667$ (*2*). Points I–VIII show the results for modes I–VIII, respectively

band $0 \le c_r \le 1$. This means that the development of such disturbances is caused by the existence of critical layers in the flow domain, where the phase velocity of the disturbance coincides with the flow velocity [23]. Because of intense damping of these disturbances, their properties are not considered here in detail, and we restrict ourselves to the case of the constant viscosity model. It should be noted only that the boundaries of the set S_v practically do not differ from those in the perfect gas with Satherland's viscosity law [6, 7] even if the maximum excitation of internal degrees of freedom (see Fig. 3.6) are taken into account in the case considered in the present study. For the excited gas there is a certain shift of the eigenvalues toward greater damping decrements, which is negligibly small as compared to the values of $c_i \in [0.1;\ 1]$. It follows from Figs. 3.5 and 3.6 that all specific features of the dynamics of the disturbance spectrum obtained in [6, 7] as a function of the flow parameters and wave number α are retained. In particular as the Reynolds number

Re, Mach number M, and wavenumber α increase, the Y-shaped structure of the set S_v transforms to the H-shaped structure. At moderate values of the wave number α, the triple splitting of the V-shaped head part of the Y-shaped structure noted in [6] is retained. It should be noted that this splitting is explained in [6] by the presence of three dissipative terms with different coefficients in a system of equations for a perfect gas similar to Eq. (3.15). According to [6], splitting in the present case can be expected to induce the formation of the fourth branch caused by the presence of the dissipative term in the equation for the disturbances of vibrational temperature. However, such a branch is not obtained in the calculations performed in this study (see, e.g., Fig. 3.6a).

At moderate supersonic Mach numbers the set S_a is presented in Fig. 3.5 by two eigenvalues corresponding to modes I and II, which become closer to neutral modes with $c_i \sim 10^{-5}$ as the Reynolds number increases (cf. [25]). It is seen in Fig. 3.6 that higher modes III, V, VII, etc. and IV, VI, VIII, etc., appear with increasing Mach number and wavenumber α; these modes are localized near the axis $c_i = 0$. In accordance with Fig. 3.4 these modes approach the band $0 \leq c_r \leq 1$, and modes I, II, and IV enter this band, where the necessary condition for instability development is satisfied for the inviscid limits of these modes by virtue of the first Rayleigh theorem (see Sect. 3.2.2). It is seen that excitation of internal degrees of freedom of molecules leads to visible changes in the growth rates of even and odd modes from the set S_a toward enhancement of their stability.

3.3.3 Neutral Stability Contours and Critical Reynolds Numbers

The behavior of the imaginary components of the frequencies $\omega_i(\alpha) = \alpha c_i$ of modes I and II is illustrated in Fig. 3.7 for the constant viscosity model where data for perfect and vibrationally excited gases are presented. In Fig. 3.8 a comparison of the decrements (increments) in modes I and II for two models of viscosity is given for a perfect gas only. Figure 3.8 shows that stratification shifts the increments of modes I and II compared with the alternative model to the short-wave disturbances with a slight decrease of their maximum values. Influence of the vibrational excitation on these most unstable modes in a stratified flow can be estimated from the graphs of Fig. 3.9. Graphs in Figs. 3.7, 3.8, and 3.9 show that modes I and II become strongly unstable in a rather narrow interval of the wavenumber α at $M = 3$ for both viscosity models. It should be noted that mode I in the inviscid case remains stable for all Mach numbers and all values of α (see Sect. 3.2.4 and [6]). Mode II, vice versa, is unstable for the inviscid flow at $c_r > 0$, and vibrational excitation stabilizes it (see Sect. 3.2.4). The maximum growth rate takes place for mode II at $M = 5$ as shown in Figs. 3.7d and 3.8b. The peak value is greater by more that a factor of 1.5 than the maximum value in the inviscid case (see Table 3.1). It is obvious the emergence of peaks is caused by the destabilizing influence of viscosity in a narrow range of

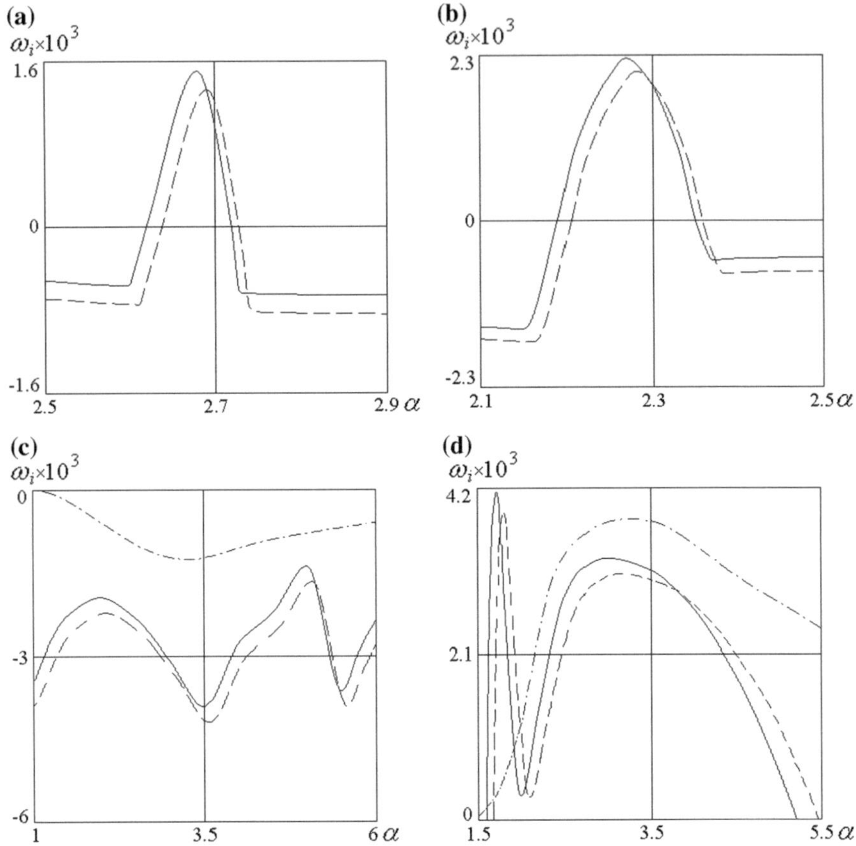

Fig. 3.7 Dependencies $\omega_i(\alpha)$ for constant viscosity model at various of the Mach numbers M and Re $= 5 \times 10^5$. **a** and **b** are M $= 3$. **c** and **d** are M $= 5$. **a** and **c** are mode I. **b** and **d** are mode II. The curves show the data for a perfect gas (*solid curves*), vibrationally excited gas with $\alpha_1 = 2$ and $\gamma_v = 0.667$ (*dashed curves*), and ideal gas with $\gamma_v = 0$ (*dot-and-dashed curves*)

wavenumbers α. With a further increase in the Mach number, mode I becomes stable (see Figs. 3.7c and 3.8a).

At the same time mode II remains unstable in the entire examined range M $\in$ [3; 15] for both viscosity models. As a whole, the calculations show that viscosity is a stabilizing factor for both modes and viscosity models in the major part of the domains of variation of Re, M, and α.

Influence of the vibrational excitation on most unstable acoustic modes of disturbances for considered models can be estimated from the graphs of Figs. 3.7 and 3.9. It can be seen that the excitation leads to a decrease in the growth rates in the main part of the instability range. The relative value of the suppression of disturbances is the same for both models of viscosity.

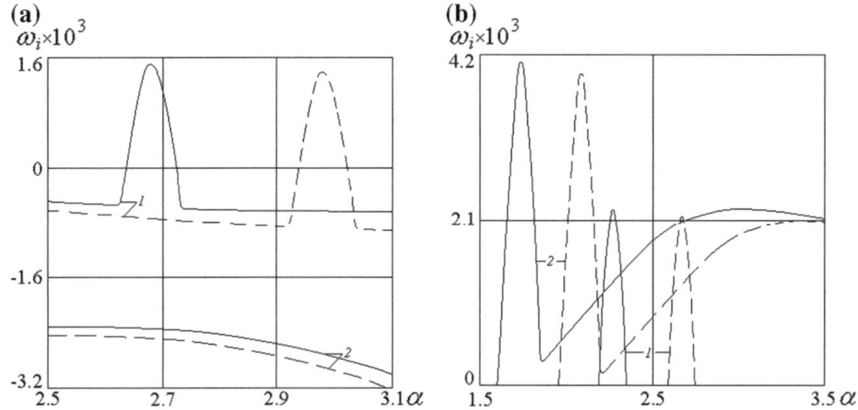

Fig. 3.8 Dependencies $\omega_i(\alpha)$ for a perfect gas at Re $= 5 \times 10^5$, M $= 3$ (*1*) and 5 (*2*). **a** is mode I. **b** is mode II. The *solid curves* are the constant viscosity model. The *dashed curves* are Sutherland's model (3.10)

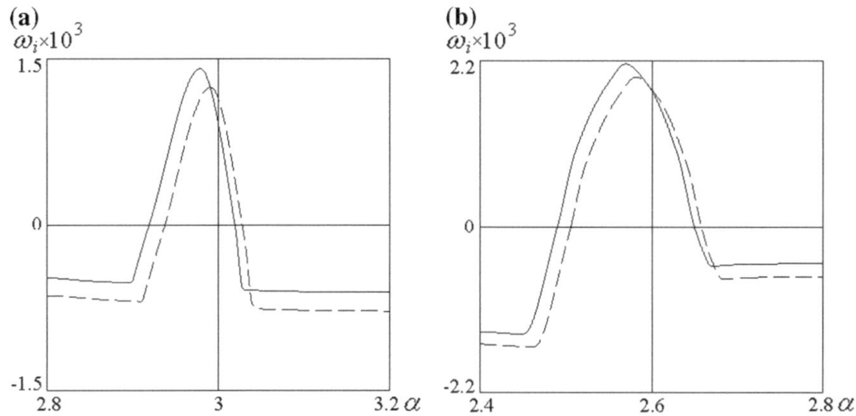

Fig. 3.9 Dependencies of $\omega_i(\alpha)$ for Sutherland's model (3.10) at Re $= 5 \times 10^5$, M $= 3$. **a** is mode I. **b** is mode II. The *solid curve* is a perfect gas. The *dashed curve* is the vibrationally excited gas with $\alpha_1 = 2$ and $\gamma_v = 0.667$

Figure 3.10 shows the neutral stability curves of modes I and II for both models viscosity at M $= 3$. The dashed lines correspond to the limiting excitation of the vibrational mode at $\gamma_v = 0.667$. These curves can be considered as a shift of the corresponding solid curves for a nonexcited perfect gas to the large numbers Re and α without changing their form. In particular, we observe absence of deformation of the internal instability domain. It is seen that mode II is less stable than mode I over the entire range of parameters. It follows from comparing the critical values of the Reynolds numbers Re_{cr} denoted by points K_2, K_2' and K_1, K_1', respectively.

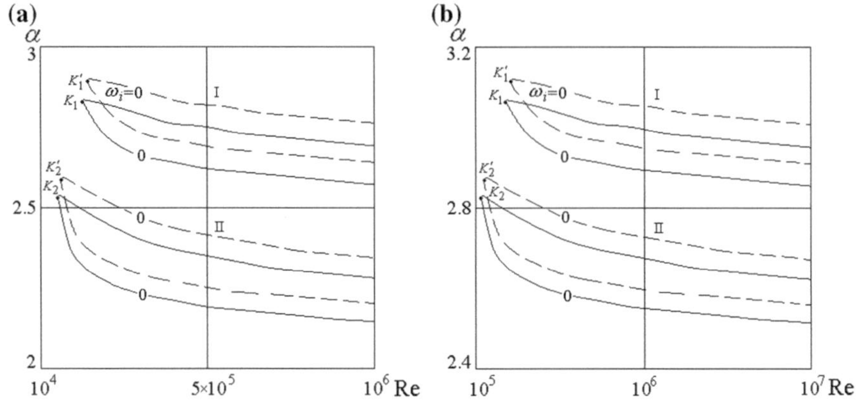

Fig. 3.10 Neutral stability curves $\omega_i(\text{Re}, \alpha) = 0$ for two models of viscosity at M = 3. **a** is the constant viscosity model. **b** is the Sutherland's model (3.10). The *solid* and *dashed curves* show the results for a perfect gas and for a vibrationally excited gas with $\alpha_1 = 2$ and $\gamma_v = 0.667$, respectively. The data for modes I and II are marked by I and II, respectively. K_1 and K_1' are the critical points for mode I. K_2 and K_2' are the critical points for mode II

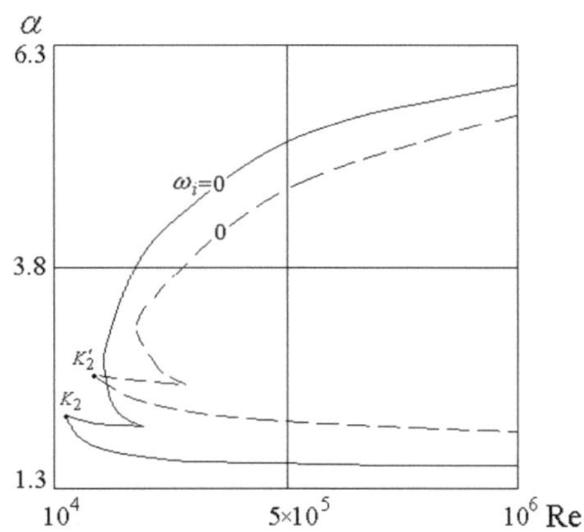

Fig. 3.11 The neutral stability curves $\omega_i(\text{Re}, \alpha) = 0$ of the mode II for a perfect gas at M = 5. The *dashed* and *solid curves* show the results for the models of Sutherland (3.10) and constant viscosity, respectively. K_2 and K_2' are the critical points of mode II

The neutral curves for two models of viscosity are additionally compared in Fig. 3.11 at Mach number M = 5 for the perfect gas. Analysis of the graphs in Figs. 3.10 and 3.11 allows us to conclude that along a whole range of Reynolds numbers the effect of viscosity stratification is mainly reduced to increasing the dissipative effect. This is manifested in narrowing of the instability area and in essentially increasing the critical Reynolds numbers. At the same time we can note that the characteristic features of the behavior of neutral stability curves in the long-distance supersonic region for $M \geq 3$ don't change for either model. In particular for $M \geq 5$ the instability domain is expanded with increasing Reynolds number (see Fig. 3.11). Furthermore, in both cases the curves have two minima of the Reynolds number. The sharp peak close to $\alpha = 1.5$ corresponds to the sharp peak of the growth

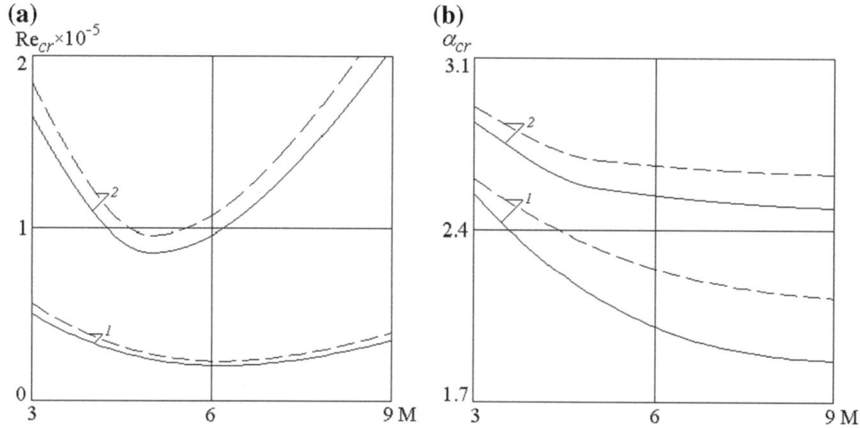

Fig. 3.12 Dependencies of Re_{cr} (M) and α_{cr} (M) for constant viscosity model (*1*) and Sutherland's model (3.10) (*2*). **a** shows dependencies Re_{cr} (M). **b** shows dependencies α_{cr} (M). The *solid* and *dashed curves* show the results for a perfect gas and for a vibrationally excited gas with $\alpha_1 = 2$ and $\gamma_v = 0.667$, respectively

rate in the graphs of the Figs. 3.7, 3.8, and 3.9 which is associated with additional viscous instability. On the contrary the minimum on a gently sloping part of the curves is determined by dissipative influence of viscosity on the maximum growth rate of inertial instability at $M = 5$.

The curves on Fig. 3.12 give one a possibility to qualitatively estimate the effect of the viscosity model, compressibility, and vibrational excitation on the critical parameters of mode II. "Viscous" stratification increases the value of Re_{cr} by several times as compared to the case of a constant viscosity for a perfect gas in whole range of the Mach numbers. At the same time the wavelengths of the most unstable disturbances (the critical wavenumbers α_{cr} (M)) change insignificantly with transition from one model to the other. The excitation of the vibrational mode increases the critical Reynolds numbers and values of the critical wavenumbers for both models under all Mach numbers but this effect is significantly less than influence of stratification. It is seen that the nonmonotonicity of all curves depending on the Mach number has an identical character. Indeed, at moderate supersonic values of the Mach numbers $M \leq 6$, stability of flow decreases with increasing compressibility. On the other hand, further increase of compressibility (at Mach numbers $M > 6$) leads to a stabilization of the flow. The ambiguous effect of the Mach number on stability is not unique in this case; it is also observed, e.g., in a supersonic boundary layer [26].

Table 3.2 presents exact quantity estimations of the influence of each considered factor on the critical Reynolds numbers and critical wavenumbers of mode II. Particularly, taking into account Sutherland's law of temperature dependence of the viscosity (3.10) increases Re_{cr} from three to six times at $3 \leq M \leq 15$. At the same time α_{cr} increases within the limits 30%. The relative increase in the numbers Re_{cr}

Table 3.2 Critical values of the Reynolds number Re_{cr} and wavenumbers α_{cr} for mode II

M	A perfect gas ($\alpha_1 = 0, \gamma_v = 0$)				The vibrationally excited gas ($\alpha_1 = 2, \gamma_v = 0.667$)			
	The constant viscosity model		The Sutherland's model (3.10)		The constant viscosity model		The Sutherland's model (3.10)	
	Re_{cr}	α_{cr}	Re_{cr}	α_{cr}	Re_{cr}	α_{cr}	Re_{cr}	α_{cr}
3	50060	2.5460	164870	2.8398	56060	2.6039	184190	2.9025
5	23830	2.1310	85720	2.5688	26930	2.3377	95750	2.6831
7	21640	1.9301	121830	2.5156	24230	2.1801	136080	2.6435
9	35080	1.8706	205110	2.4884	39630	2.1226	229110	2.6241
11	55750	1.8790	308170	2.4851	62720	2.1144	344230	2.6218
13	75240	1.8840	461550	2.4872	84940	2.1119	515550	2.6255
15	85150	1.8110	655850	2.4950	96150	2.0660	732580	2.6394

under excitation of the vibrational mode practically coincides for two models of viscosity and reaches approximately 10–12%.

The following conclusions can be drawn from the present study.

Similar to the case of a perfect gas, the set of eigenvalues of the linear problem of stability of the supersonic viscous Couette flow of a thermally excited molecular gas is divided into two uninteresting sets S_a and S_v. As $\mathrm{Re} \to \infty$, the eigenvalues from the set S_a converge to the eigenvalues for inviscid acoustic modes I, II, III, etc. The eigenvalues from the set S_v have no asymptotic relationship with the inviscid problem and are characterized by large damping decrements.

Viscous modes I and II of the set S_a are the most unstable modes. Viscosity produces an ambiguous effect on the behavior of these modes. Mode I, whose inviscid limit is stable in the entire range of Mach numbers and wavenumbers, becomes unstable at finite Reynolds numbers $\mathrm{Re} > 3 \times 10^4$ in a narrow domain $2.5 < \alpha < 2.75$ and $\mathrm{M} \approx 3$. At other values of these parameters, viscosity leads to enhancement of stability of mode I as compared to the inviscid flow. Viscous mode II, which is unstable in the inviscid flow for all values of M and α for which $0 < c_r < 1$, becomes stable at $\mathrm{M} < 3$ and finite Reynolds numbers. At $\mathrm{M} \geq 3$ the growth rate has a peak in a certain small range of wavenumbers, and the magnitude of this peak is appreciably greater than the corresponding maximum value for the inviscid flow. Outside this range of α, mode II becomes completely stabilized at $\mathrm{Re} < \mathrm{Re}_{cr}$ in the viscous case. Nevertheless, there is a domain of instability of mode II on the plane of parameters (Re, α) at $\mathrm{M} \geq 3$. The "viscous" stratification of the Couette flow when using Sutherland's model greatly enhances the stabilization of flow. At $3 \leq \mathrm{M} \leq 15$ the corresponding critical Reynolds numbers of the most unstable mode II are 3–6 times higher as compared to the model of constant viscosity.

At the same time the relative decrease in growth rates of viscous modes I and II under vibrational excitation is practically equal for both models of viscosity. The increase in the critical Reynolds numbers is within the limits of 10–12% in both cases.

An important conclusion implies that a simple model of constant viscosity qualitatively reproduces all the characteristic features of the development of viscous disturbances of acoustic modes I and II in a perfect gas and in a vibrationally excited gas. Particularly, there is a narrow range of wavenumbers near the Mach number $\mathrm{M} \approx 3$ where destabilization of mode I which is stable in the inviscid limit takes place. Also a significant increase in the growth rate of mode II compared to the inviscid limit at $\alpha \approx 1.5$ is observed for the supersonic range $5 \leq \mathrm{M} \leq 15$. The neutral stability curves exhibit typical features of their behavior, which, in particular, includes broadening of the instability domain with the growth of Re and occurrence of two maximums on the upper branch. The nonmonotonic dependence of the critical Reynolds number Re_{cr} on the Mach number M is the same for all cases considered in this study. This conclusion allows us to use the simple model of the constant viscosity in the following chapters of the book.

References

1. Drazin, P.G., Reid, W.H.: Hydrodynamic Stability. Cambridge University Press, Cambridge (2004)
2. Romanov, V.A.: Stability of a plane-parallel Couette flow. Dokl. Akad. Nauk SSSR **196**, 1049–1051 (1971) (in Russian)
3. Gol'dshtik, M.A., Shtern, V.N.: Hydrodynamic Stability and Turbulence. Nauka, Novosibirsk (1977) (in Russian)
4. Joseph, D.D.: Stability of Fluid Motion. Springer, Berlin (1976)
5. Hanifi, A., Henningson, D.S.: The compressible inviscid algebraic instability for streamwise independent disturbances. Phys. Fluids **10**, 1784–1786 (1998)
6. Duck, P.W., Erlebacher, G., Hussaini, M.Y.: On the linear stability of compressible plane Couette flow. J. Fluid Mech. **258**, 131–165 (1994)
7. Hu, S., Zhong, X.: Linear stability of viscous supersonic plane Couette flow. Phys. Fluids **10**, 709–729 (1998)
8. Malik, M., Dey, J., Alam, M.: Linear stability, transient energy growth, and the role of viscosity stratification in compressible plane Couette flow. Phys. Rev. E **77**, 036322(15) (2008)
9. Lin, C.C.: The Theory of Hydrodynamic Stability. University Press, New York (1966)
10. Nagnibeda, E.A., Kustova, E.V.: Non-equilibrium Reacting Gas Flows. Kinetic Theory of Transport and Relaxation Processes. Springer, Berlin (2009)
11. Grigor'ev, Yu.N., Ershov, I.V.: Stability of Flows of Relaxing Molecular Gases. Izd. Sib. Otd. Ross. Akad. Nauk, Novosibirsk (2012) (in Russian)
12. Ferziger, J.H., Kaper, H.G.: Mathematical Theory of Transport Processes in Gases. North Holland Publ. Comp, Amsterdam (1972)
13. Grigor'ev, Yu.N., Yershov, I.V.: Linear stability of an inviscid shear flow of a vibrationally excited diatomic Gas. J. Appl. Math. Mech. **75**, 410–418 (2011)
14. Blumen, W.: Shear layer instability of an inviscid compressibled fluid. J. Fluid Mech. **40**, 769–781 (1970)
15. Drazin, P.G., Howard, L.N.: Hydrodynamic stability of parallel flow of inviscid fluid. In: Chernyi, G.G. et al. (Eds.) Advance in Applied Mechanics, vol. 9, pp. 1–89. Academic Press, New York (1996)
16. Howard, L.N.: Note on a paper of John W. Miles. J. Fluid Mech. **10**, 509–512 (1961)
17. Canuto, C., Hussaini, M.Y., Quarteroni, A., Zang, T.A.: Spectral Methods in Fluid Dynamics. Springer, Berlin (1988)
18. Trefethen, L.N.: Spectral Methods in Matlab. Society for Industrial and Applied Mathematics. SIAM, Philadelphia (2000)
19. Grigor'ev, Yu.N., Ershov, I.V.: Energy estimate of the critical Reynolds numbers in a compressible Couette flow. Effect of bulk viscosity. J. Appl. Mech. Tech. Phys. **51**, 669–675 (2010)
20. Grigor'ev, Yu.N., Ershov, I.V.: Critical Reynolds number of the Couette flow in a vibrationally excited diatomic gas. Energy approach. J. Appl. Mech. Tech. Phys. **53**, 517–531 (2012)
21. Korn, G.A., Korn, T.M.: Mathematical Handbook for Scientists and Engineers. McGraw-Hill, New York (1961)
22. Michalke, A.: On the inviscid instability of the hyperbolic tangent velocity profile. J. Fluid Mech. **19**, 543–556 (1964)
23. Morawetz, C.S.: The eigenvalues of some stability problems involving viscosity. J. Rat. Mech. Anal. **1**, 79–603 (1952)
24. Mack, L.M.: On the inviscid acoustic-mode instability of supersonic shear flows. Part I: two-dimensional waves. Theor. Comput. Fluid Dyn. **2**, 97–123 (1990)
25. Grigor'ev, Yu.N., Ershov, I.V.: Linear stability of the Couette flow of a vibrationally excited gas. 1. Inviscid problem. J. Appl. Mech. Tech. Phys. **55**, 258–269 (2014)
26. Gaponov, S.A., Maslov, A.A.: Development of Perturbations in Compressible Flows. Nauka, Novosibirsk (1980) (in Russian)

Chapter 4
Asymptotic Theory of Neutral Linear Stability Contours in Plane Shear Flows of a Vibrationally Excited Gas

Abstract An asymptotic theory of the neutral stability curve for plane shear flows of a vibrationally excited gas is developed in the chapter. The initial mathematical model consists of equations of two-temperature viscous gas dynamics, which are used to derive a spectral problem for a linear system of eighth-order ordinary differential equations. Unified transformations of the system for all shear flows are performed in accordance with the classical scheme. The spectral problem for the supersonic plane Couette flow with two boundary conditions, which was not considered previously even for perfect gas, is reduced to an algebraic secular equation with separation into the "inviscid" and "viscous" parts. The neutral stability curves obtained on the basis of the numerical solution of the secular equation agree well with the previously obtained results of the direct numerical solution of the original spectral problem.

The classical investigations of Helmholtz, Heisenberg, Kelvin, Rayleigh, and Sommerfeld on hydrodynamic stability theory were exclusively based on analytical asymptotical methods. These methods were practically the only available tools when the theory was developed in the 1940–1950s. With use of these methods, the first meaningful results were obtained, mainly for the boundary layer on a flat plate. These results were summarized in monograph [1]. Further improvement of analytical approaches made it possible to find some general features in the field of hydrodynamic stability and transition to turbulence, and also to estimate the critical parameters of some flows important for practice, which were fairly close to actually observed values. At the same time owing to their inherent nature, asymptotic methods have some constraints, first of all, on the flow parameter values, i.e., Mach and Reynolds numbers, phase velocities, and disturbance wavelengths. Moreover, activities based on asymptotic methods often involved a number of assumptions whose validity could not be verified at that time.

As computer hardware and software were developed, they provided a possibility of the direct numerical solution of problems of the linear stability theory for flows important in practice without constraints on flow parameters. The founder of the new research direction started in the 1960 s was I.M. Mack; multiple references to his works can be found, e.g., in monograph [2]. These and subsequent numerical studies of other authors, including recent investigations, shifted the use of asymptotic

Y.N. Grigoryev and I.V. Ershov, *Stability and Suppression of Turbulence in Relaxing Molecular Gas Flows*, Fluid Mechanics and Its Applications 117,
DOI 10.1007/978-3-319-55360-3_4

approximations to the background. However, these new investigations also assisted in improvement of the classical analytical methods, particularly, in refining the area of their applicability and avoiding unjustified assumptions. Therefore, verified analytical results are described in monographs on the hydrodynamic stability theory and are still used for comparisons with results of numerical and experimental studies. New problems of hydrodynamic stability, e.g., for flows of optically active or thermally nonequilibrium molecular gases [3] stimulate generalization of the classical results [1], first of all, approximate estimates of the critical Reynolds numbers, to media and flows that were not considered in earlier researches.

This chapter describes the asymptotic theory of the neutral stability curve for plane shear flows of a vibrationally excited gas. For simplicity, in this chapter the study is confined to the case of temperature-independent transport coefficients.

In Sect. 4.1 the main stages of derivation of the secular (characteristic) equation are presented.These transformations are the same for all shear flows. Here we follow the scheme first described in [1] (see also [4]). The asymptotic expansion of the original system of linearized equations with respect to the small parameter $1/\mathrm{Re}$ reveals "inviscid" and "viscous" linearly independent solutions.

In Sect. 4.1.1 two linearly independent "inviscid" solutions for $\mathrm{Re} \to \infty$ are constructed by the Frobenius method on the basis of the asymptotic expansion in the vicinity of the critical layer. The calculations deal with the equation for the pressure perturbation.

In Sect. 4.1.2 the original system for finding the "viscous" solutions is subjected to some simplification and converted to a sixth-order system similar to the Dunn-Lin system [1]. The momentum and energy (static temperature) equations involved in this system are reduced to the Airy equations. As a result the "viscous" solutions are presented via the generalized Airy functions of the zeroth, first, and second orders. The transformations of the system and all further discussion in Sect. 4.2 are performed for a supersonic plane Couette flow of a vibrationally excited gas with a linear velocity and parabolic static temperature profiles.

In Sect. 4.2.1 the secular equation is derived from boundary conditions for a sixth-order system of ordinary differential equations simplified on the basis of the asymptotic properties of the Airy functions. The use of the Airy functions makes it possible to express the "viscous" part of the secular equation via the tabulated Tietjens function and its derivative, which are usually used in asymptotic stability theory [1, 2].

In Sect. 4.2.2 the secular equation is used to derive a system for asymptotic estimation of the critical Reynolds numbers. Relations for intermediate asymptotic solutions for the upper and lower branches of the neutral stability curve are also obtained.

Section 4.2.3 describes the numerical calculations of the neutral stability curves on the basis of the secular equation. Comparisons show that the calculations and asymptotic formulas of Sect. 4.2.2 are in good agreement with the results of the direct numerical solution of the original spectral problem, which were reported in Chap. 3. An increase in the level of excitation leads to the growth of the critical Reynolds number and to shifting of the neutral stability curve toward higher wave numbers.

4.1 Asymptotic Solutions of Linear Stability Equations

As in Chap. 3, periodic perturbations along x are considered in the form of travelling plane waves:

$$\boldsymbol{q}(x,\, y,\, t) = \boldsymbol{q}(y)\, \exp\left[i\alpha(x - ct)\right],$$

$$\boldsymbol{q}(x,\, y,\, t) = (\hat{u}_x,\, \hat{u}_y,\, \hat{\rho},\, \hat{T},\, \hat{T}_v,\, \hat{p}), \quad \boldsymbol{q}(y) = (u,\, \alpha v,\, \rho,\, \theta,\, \theta_v,\, p)\,.$$

Here α is the wavenumber along the periodic variable x, $c = c_r + ic_i$ is the complex phase velocity, i is the imaginary unity, and $\boldsymbol{q}(y)$ is the perturbation amplitude vector. Substitution of $\boldsymbol{q}(x,\, y,\, t)$ into the linearized system from Chap. 3 with allowance for constant values of the transport coefficients yields the following equations for perturbation amplitudes (cf. Chap. 3):

$$D\rho + \alpha\rho_s' v + \rho_s\sigma = 0, \tag{4.1}$$

$$\frac{1}{\mathrm{Re}}\triangle u - \rho_s Du - \alpha\rho_s v U_s' - i\alpha\varepsilon = 0, \tag{4.2}$$

$$\frac{\alpha}{\mathrm{Re}}\triangle v - \alpha\rho_s Dv - \varepsilon\,' = 0, \tag{4.3}$$

$$\frac{\gamma}{\mathrm{RePr}}\triangle\theta - \rho_s D\theta - \alpha\rho_s v T_s' - (\gamma - 1)\sigma +$$

$$\frac{2\gamma\,(\gamma - 1)\,\mathrm{M}^2}{\mathrm{Re}}\left(u' + i\alpha^2 v\right) U_s' + \frac{\gamma_v\rho_s}{\tau}(\theta_v - \theta) = 0, \tag{4.4}$$

$$\frac{20\gamma\gamma_v}{33\mathrm{RePr}}\triangle\theta_v - \gamma_v\rho_s D\theta_v - \alpha\gamma_v\rho_s v T_s' - \frac{\gamma_v\rho_s}{\tau}(\theta_v - \theta) = 0, \tag{4.5}$$

$$\gamma\mathrm{M}^2 p = \rho_s\theta + \rho T_s. \tag{4.6}$$

Here

$$D = i\alpha(U_s - c), \quad \sigma = \alpha(v' + iu), \quad \varepsilon = p - \frac{\sigma}{\mathrm{Re}}\left(\alpha_1 + \frac{1}{3}\right), \quad \triangle = \frac{d^2}{dy^2} - \alpha^2,$$

and the primes hereinafter mean differentiation with respect to the variable y. The system of linear equations (4.1)–(4.6) has the eighth order and, combined with the homogeneous boundary conditions

$$u\big|_{y=0} = u\big|_{y=1} = v\big|_{y=0} = v\big|_{y=1} = \theta\big|_{y=0} = \theta\big|_{y=1} = \theta_v\big|_{y=0} = \theta_v\big|_{y=1} = 0, \tag{4.7}$$

defines the spectral problem whose eigenvalues are the complex phase velocity of perturbations c, whereas Reynolds number Re, the Mach number M, and the wavenumber α serve as parameters. The asymptotic solutions of the system for high Reynolds numbers are constructed in the form of the series of the perturbation theory as

$$\boldsymbol{q}(y) = \boldsymbol{q}_0(y) + \frac{1}{\text{Re}}\boldsymbol{q}_1(y) + \cdots .$$

In the zeroth approximation we obtain a system of equations for inviscid perturbations, which was considered in Chap. 2. In particular it was demonstrated that it reduces to a second-order linear equation for the pressure perturbation. Therefore, the zeroth approximation allows one to find only two linearly independent solutions. The remaining six solutions have to be found from direct consideration of system (4.1)–(4.6).

4.1.1 *Asymptotics of Inviscid Solutions in Neighborhood of a Singular Point*

The general qualitative properties of the equation for the pressure perturbation were studied in Chap. 2. Here it is convenient to consider this equation in a non-self-conjugate form (cf. Chap. 2)

$$p'' - \frac{T_s}{W^2}\left(\frac{W^2}{T_s}\right)' p' - \alpha^2\left(1 - \frac{\text{M}^{*2}\,W^2}{T_s}\right)p = 0, \tag{4.8}$$

where

$$W = U_s - c, \quad \text{M}^{*2} = m^2\text{M}^2, \quad m^2 = m_r^2 + i\,m_i^2,$$

$$m_r^2 = \frac{R_1\,(1 + \gamma_v + \alpha\tau c_i) + \Delta^2}{R_1^2 + \Delta^2}, \quad m_i^2 = -\frac{\gamma_v}{\gamma}\,\frac{(\gamma - 1)\Delta}{R_1^2 + \Delta^2},$$

$$R_1 = 1 + (\gamma_v/\gamma) + \alpha\tau\, c_i, \quad \Delta = \alpha\tau\,(U_s - c_r)\,.$$

For the neutral perturbations considered below, we have $c \equiv c_r$. As the first Rayleigh condition $c_r \in [0;\ 1]$ has also to be satisfied for neutral perturbations (Chap. 2), then $y = y_c$ is a regular singular point of the considered equation. Its solutions in the neighborhood of the singular point are found by the Frobenius method [5]. In what follows the subscript "c" is used to indicate the values of the variables at the point y_c, where the phase velocity is equal to the free-stream velocity (in the critical layer): $U_s(y_c) = c$. In particular for the linear velocity profile where $U_s = y$ it means $y_c = c$.

Let us introduce a new independent variable, $\eta = y - y_c$. Using this variable, it is convenient to present Eq. (4.8) in the form

$$p'' - a(\eta)p' - \alpha^2 b(\eta)p = 0\,,$$

$$a(\eta) = \frac{T_s}{\eta^2}\left(\frac{\eta^2}{T_s}\right)', \quad b(\eta) = 1 - \frac{\mathrm{M}^{*2}\,\eta^2}{T_s}\,. \tag{4.9}$$

The solution of Eq. (4.9) is sought in the form of the generalized power-law series [5]:

$$p(\eta) = \eta^{\rho} \sum_{k=0}^{\infty} c_k\, \eta^k, \quad c_0 \neq 0. \tag{4.10}$$

The coefficients of Eq. (4.9) are preliminary expanded into power-law series with respect to the variable η in the neighborhood of the singular point y_c ($\eta = 0$):

$$a(\eta) = 2\eta + A\eta^2 + \left(A^2 + 2B\right)\eta^3 + \cdots,$$

$$b(\eta) = 1 - \frac{m_{rc}^2\,\mathrm{M}^2}{T_{sc}}\,\eta^2 + \frac{m_{rc}^2\,\mathrm{M}^2\,T'_{sc}}{T_{sc}^2}\,\eta^3 + \cdots,$$

$$A = \frac{U''_{Sc}}{U'_{Sc}}\frac{T'_{sc}}{T_{sc}}, \quad B = \frac{U''^2_{Sc}}{4U'^2_{Sc}} + \frac{U'''Sc}{3U'Sc} - \frac{T''_{sc}}{2T_{sc}}.$$

Then substitute Eq. (4.10) into (4.9), take account of the expansions obtained, and equate the values of the coefficients at each power of η^k to zero. The coefficient at the lowest power of $\eta^{\rho-2}$ yields the following governing equation for ρ [5]:

$$\rho\,(\rho - 1)\,c_0 - 2\rho c_0 = 0. \tag{4.11}$$

As $c_0 \neq 0$, it follows that Eq. (4.11) has two roots $\rho_1 = 3$ and $\rho_2 = 0$. As $\rho_1 \neq \rho_2$ and $\rho_1 - \rho_2 = 3$ is an integer number, then, according to [5], two linearly independent solutions of Eq. (4.9) are constructed in the form

$$p_1(\eta) = \eta^3 \sum_{k=0}^{\infty} c_k^{(1)}\,\eta^k, \quad p_2(\eta) = C\,p_1(\eta)\ln(\eta) + \sum_{k=0}^{\infty} c_k^{(2)}\,\eta^k.$$

Without loss of generality one can assume that $c_0 = 1$. Substitution of $p_1(\eta)$ with $\rho_1 = 3$ into Eq. (4.9) and equating the coefficients at consecutive powers of η^k to zero allows us to obtain recurrent expressions for the coefficients $\left\{c_k^{(1)}\right\}$. As a result, we have

$$p_1(\eta) = \eta^3 + \frac{3A}{4}\eta^4 + \frac{6(A^2 + 2B) + \alpha^2}{10}\eta^5 + \cdots.$$

Substituting $p_2(\eta)$ into Eq. (4.9), we can separate the singular part and obtain recurrent expressions for the unknown coefficients $\left\{c_k^{(2)}\right\}$. As a result, the second solution is written in the form[1]

$$p_2(\eta) = 1 - \frac{\alpha^2}{6} A p_1(\eta)\ln(\eta) - \frac{\alpha^2}{2}\eta^2 -$$

$$\frac{\alpha^2}{12}\left(2B + \frac{\alpha^2}{2} + \frac{A^2}{12} + \frac{m_{rc}^2 \mathrm{M}^2}{T_{sc}}\right)\eta^4 + \cdots, \quad \eta > 0;$$

$$p_2(\eta) = 1 - \frac{\alpha^2}{6} A p_1(\eta)\left(\ln(\eta) - i\pi\right) - \frac{\alpha^2}{2}\eta^2 -$$

$$\frac{\alpha^2}{12}\left(2B + \frac{\alpha^2}{2} + \frac{A^2}{12} + \frac{m_{rc}^2 \mathrm{M}^2}{T_{sc}}\right)\eta^4 + \cdots, \quad \eta < 0.$$

In what follows we will need expressions for the remaining "inviscid" solutions for the linear velocity profile. They are expressed via the pressure perturbation as follows ($k = 1, 2$):

$$u_{\mathrm{k}i}(\eta) = -\frac{T_s}{\alpha\,\eta}\left(\frac{p_k'}{\alpha\,\eta} + \alpha\, p_k\right), \quad v_{\mathrm{k}i}(\eta) = \frac{i p_k' T_s}{\alpha^2\,\eta},$$

$$\theta_{\mathrm{k}i}(\eta) = \frac{T_s}{\alpha^2\,\eta^2}\left(p_k' T_s' + (\gamma - 1)\,\alpha^2\,\eta^2\,\mathrm{M}^2 p_k\left[1 - \frac{\tau\gamma_v}{\gamma + \gamma_v + i\,\gamma\tau\alpha\eta}\right]\right),$$

$$\theta_{v,\,\mathrm{k}i}(\eta) = \frac{i\,T_s}{\alpha\,\eta}\left(\frac{\alpha v_k T_s'}{T_s} - \frac{i\gamma\,(\gamma - 1)\,\tau\alpha\eta \mathrm{M}^2 p_k T_s}{\gamma + \gamma_v + i\gamma\tau\alpha\eta}\right),$$

4.1.2 *Asymptotics of Viscous Solutions at High Reynolds Numbers*

To find the "viscous" solutions we have to transform and simplify system (4.1)–(4.6), similar to [2] (see also [4]). First, let us combine the momentum equations to eliminate terms with pressure fluctuations. The equation for the perturbation of the transverse velocity v is multiplied by $i\alpha$ and then subtracted from the preliminary differentiated equation for the perturbation of the streamwise velocity u. As a result, one obtains a corollary consisting of two momentum equations in the form

$$\frac{1}{\mathrm{Re}}\left[\Delta u' - i\alpha^2 \Delta v\right] - (\rho_s D u)' - \left(\alpha\rho_s v U_s'\right)' - i\alpha^2 \rho_s D v = 0. \tag{4.12}$$

---

[1]The choice of the minus sign in the logarithm argument at $\eta < 0$ was justified in [2] by referring to W. Tollmien, who showed that this is necessary for profiles with $U_{sc}' > 0$ to ensure a correct jump of the phase of the perturbation of the streamwise velocity u across the critical layer.

It should be noted that the resultant third-order equation does not contain the term with bulk viscosity (coefficient α_1). Therefore, the expected dissipative effect in linear theory is determined only by relaxation of the vibrational mode.

The approach to estimating the terms in system (4.1)–(4.6) for the case of the boundary layer based on the relationship of the orders of a small parameter introduced in a special manner and used in [4, 6] seems to be somewhat artificial. Therefore, we simplify Eq. (4.12) similar to [2] (W. Tollmien), leaving only the leading terms, as is usually done in mathematical papers on the spectral theory [7, 8]. The form of the remaining equations can be taken from [4], retaining the relaxation terms in the equations for the temperatures, θ and θ_v. Thus, one obtains the following system:

$$v' + iu - i\left(U_s' - c\right)\frac{\theta}{T_s} = 0, \tag{4.13}$$

$$u''' - i\alpha\mathrm{Re}\left(U_s' - c\right)\frac{u'}{T_s} = 0, \tag{4.14}$$

$$\theta'' - i\alpha\mathrm{PrRe}\left(U_s' - c\right)\frac{\theta}{T_s} + \frac{\gamma_v\mathrm{PrRe}}{\gamma\tau\, T_s}\left(\theta_v - \theta\right) = 0, \tag{4.15}$$

$$\theta_v'' - \frac{33}{20}\, i\alpha\mathrm{PrRe}\left(U_s' - c\right)\frac{\theta_v}{T_s} - \frac{33\mathrm{PrRe}}{20\gamma^2\tau\, T_s}\left(\theta_v - \theta\right) = 0, \tag{4.16}$$

It should be noted that the thus-simplified system has the eighth order, like the original system. If we now pass to a perfect gas at $\theta_v = \gamma_v = 0$, then the system coincides with the Dunn-Lin “viscous” system [4].

Below all calculations are performed for a supersonic Couette flow for which the equilibrium steady-state solution of the original system of equations is valid:

$$U_s(y) = y, \quad T_s(y) = T_{v,s}(y) = 1 + \frac{(\gamma - 1)\mathrm{PrM}^2}{2}(1 - y^2),$$

$$\rho_s(y) = \frac{1}{T_s(y)}, \quad p_s(y) = \frac{1}{\gamma\mathrm{M}^2}, \tag{4.17}$$

(see also (3.11)) which satisfies the boundary conditions

$$U_s(0) = 0, \quad U_s(1) = 1, \quad T_s(1) = 1, \quad \left.\frac{dT_s}{dy}\right|_{y=0} = 0. \tag{4.18}$$

Nevertheless, all principal moments of subsequent transformations have a universal character.

The momentum equation (4.14) is transformed in the following way. Introduce new independent [9] and dependent [4] variables:

$$Y = \left[\frac{3}{2} \int\limits_0^y \sqrt{\frac{t-c}{T_s}}\, dt \right]^{2/3}, \quad U(Y) = u' \sqrt{\frac{dY}{dy}}. \tag{4.19}$$

Using them to express the derivatives involved into Eq. (4.14), one derives

$$u'(y) = U(Y) \left(\frac{dY}{dy} \right)^{-1/2},$$

$$u'''(y) = \frac{d^2U}{dy^2} \left(\frac{dY}{dy} \right)^{3/2} - \frac{U(Y)}{2} \left(\frac{dY}{dy} \right)^{3/2} \left[\frac{d^3Y}{dy^3} - \frac{3}{2} \left(\frac{d^2Y}{dy^2} \right)^2 \left(\frac{dY}{dy} \right)^{-1} \right].$$

Substitution of these expressions into Eq. (4.14) yields

$$\frac{d^2U}{dY^2} \left(\frac{dY}{dy} \right)^{3/2} - U \left[i\alpha \mathrm{Re} \frac{(y-c)}{T_s} \left(\frac{dY}{dy} \right)^{1/2} + P(Y) \right] = 0, \tag{4.20}$$

where

$$P(Y) = \frac{1}{2} \left(\frac{dY}{dy} \right)^{3/2} \left[\frac{d^3Y}{dy^3} - \frac{3}{2} \left(\frac{d^2Y}{dy^2} \right)^2 \left(\frac{dY}{dy} \right)^{-1} \right], \quad \frac{dY}{dy} = \frac{y-c}{YT_s}.$$

Assuming that $\mathrm{Re} >> 1$, we can neglect the term $P(Y)$, as was done in [4], and obtain the equation

$$\frac{d^2U}{dY^2} - i\alpha \mathrm{Re} Y U = 0.$$

Introducing a new independent variable

$$\zeta = (\alpha \mathrm{Re})^{1/3}\, Y,$$

we receive the Airy equation [10]:

$$\frac{d^2U}{d\zeta^2} - i\zeta U = 0. \tag{4.21}$$

Its linearly independent solutions can be presented in several equivalent forms, e.g., by the Airy functions of the first and second kind $Ai_{1,2}(z)$ [10]. In our case it is convenient to use the generalized Airy functions of the first and second kind $A_k(z, p)$ [2], for which a number of relations needed for the further analysis are available. In particular, the next relations are used below:

$$A_k(z, 0) = A_k(z) \equiv Ai_k(z), \quad A_k(z, -1) = Ai'_k(z), \tag{4.22}$$

$$A_k(z,\, 1) = \int\limits_{\infty}^{z} A_k(t)\, dt, \quad A_k(z,\, 2) = \int\limits_{\infty}^{z} A_{k,\,1}(t)\, dt, \tag{4.23}$$

$$A_k(z,\, p-3) - zA_k(z,\, p-1) + (p-1)A_k(z,\, p) = 0, \quad k = 1,\, 2\,. \tag{4.24}$$

In known monographs on hydrodynamic stability, e.g., [1, 2] and other works, the solutions of Eq. (4.21) are written via the Hankel functions of the first and second kind $H_{1/3}^{(1,\,2)}$. This allows us to use the tabulated Tietjens functions for calculating the neutral stability curves. For this purpose we give the following expressions [2] relating the Hankel functions to the generalized Airy functions:

$$Ai_1\left(z\, e^{i\pi/6}\right) = \frac{12^{-1/6}\, i}{2}\left(\frac{2\, z^{3/2}}{3}\right)^{1/3} H_{1/3}^{(1)}\left(\frac{2\, z^{3/2}}{3}\right),$$

$$Ai_2\left(z\, e^{i\pi/6}\right) = \frac{i}{2}\, e^{2\pi i/3}\left(\frac{2\, z^{3/2}}{3}\right)^{1/3} H_{1/3}^{(2)}\left(\frac{2\, z^{3/2}}{3}\right).$$

Taking into account the homogeneity of the linear equation (4.21), we find that two solutions can be written as

$$U_{1,\,2}\,(\zeta) = A_{1,\,2}\,(i\zeta)\,. \tag{4.25}$$

Equations (4.19) and (4.25) yield the following expression for the derivative of the streamwise velocity of viscous perturbations:

$$u'_{1,\,2}\,(y) = A_{1,\,2}\,(i\zeta)\left(\frac{dY}{dy}\right)^{-1/2}.$$

To perform integration here we should note that from transition to the independent variable ζ in Eq. (4.21) one has the following formula for the differential

$$dy = \left(\frac{dy}{dY}\right)\frac{d\zeta}{(\alpha \mathrm{Re})^{1/3}} = \frac{dY}{d\zeta}\left(\frac{dy}{dY}\right)d\zeta\,.$$

Integrating $u'_{1,\,2}\,(y)$, we obtain

$$u_{1,\,2\,\mathrm{v}}\,(y) = \int\limits_{\infty}^{\xi} \frac{dY}{d\zeta}\left(\frac{dy}{dY}\right)^{3/2} A_{1,\,2}\,(i\zeta)\; d\zeta = \frac{dY}{d\zeta}\left(\frac{dy}{dY}\right)^{3/2} A_{1,\,2}\,(i\zeta,\, 1)\;, \tag{4.26}$$

where slowly changing functions are taken out from the integral sign when passing to the last equality, similar to [4].

To find the solutions of the equations for temperature perturbations (4.15) and (4.16), we perform the following transformations. In the case of diatomic gases considered in this book, one can assume that $33/(20\gamma) \approx 1$. Summarizing and subtracting Eqs. (4.15) and (4.16), we pass from them to equations for new variables

$$\theta_+ = \theta + \theta_v \frac{\gamma_v}{\gamma}, \quad \theta_- = \theta - \theta_v.$$

Correspondingly, the reverse transition is performed by the formulas

$$\theta = \frac{\gamma\,\theta_+ - \gamma_v\,\theta_-}{\gamma + \gamma_v}, \quad \theta_v = \frac{\gamma\,(\theta_+ + \theta_-)}{\gamma + \gamma_v}. \tag{4.27}$$

For θ_+ we derive the equation

$$\theta_+'' - i\alpha \mathrm{Pr}\mathrm{Re}\,(y - c)\,\frac{\theta_+}{T_s} = 0, \tag{4.28}$$

which is solved similarly to the momentum equation (4.14). We introduce the replacement of variables (cf. Eq. (4.19)):

$$Y_\theta = \left[\frac{3}{2}\mathrm{Pr}^{1/2}\int\limits_0^y \sqrt{\frac{t-c}{T_s}}\,dt\right]^{2/3} = Y\mathrm{Pr}^{1/3}, \quad \Theta = \theta_+\sqrt{\frac{dY_\theta}{dy}},$$

which yields the equation

$$\frac{d^2\Theta}{dY_\theta^2} - i\alpha \mathrm{Re} Y_\theta \Theta = 0$$

coinciding with the equation for $U(Y)$ with accuracy up to notations. This allows us to write two linearly independent solutions of Eq. (4.28) in the form

$$\theta_{+\,1,2}\,(\zeta) = A_{1,2}\,(i\zeta_\theta)\sqrt{\frac{dy}{dY_\theta}}, \quad \zeta_\theta = (\alpha \mathrm{Re})^{1/3}\,Y_\theta = \zeta \mathrm{Rr}^{1/3}. \tag{4.29}$$

The solutions for the function θ_- are constructed in the same manner. However, to avoid complication of further transformation and simultaneously to retain (at least, partially) the effect of the relaxation terms on the viscous solutions, we confine ourselves to the solutions for the imaginary part

$$\widetilde{\theta_-} = \left(\theta_- - \overline{\theta_-}\right),$$

where the bar over the function means complex conjugation. The equation for $\widetilde{\theta_-}$ has the form

$$\widetilde{\theta_-}'' - 2i\alpha \mathrm{PrRe}\,(y-c)\,\frac{\widetilde{\theta_-}}{T_s} = 0,$$

which differs from Eq. (4.28) by a constant factor in the coefficient. Correspondingly, its solutions are

$$\widetilde{\theta}_{-1,2}(\zeta) = A_{1,2}\left(i\sqrt[3]{2}\,\zeta_\theta\right)\sqrt{\frac{dy}{dY_\theta}}, \quad \zeta_\theta = (\alpha \mathrm{Re})^{1/3}\,Y_\theta = (2\mathrm{Pr})^{1/3}\,\zeta. \tag{4.30}$$

This allows us to obtain an approximate solution for the static temperature perturbation from Eqs. (4.27), (4.29), and (4.30). This solution has the form

$$\theta_{1,2\mathrm{v}} = \frac{1}{\gamma+\gamma_v}\sqrt{\frac{dy}{dY_\theta}}\left[\gamma\,A_{1,2}\,(i\zeta_\theta) - \frac{\gamma_v}{\sqrt[3]{2}}A_{1,2}\left(i\sqrt[3]{2}\,\zeta_\theta\right)\right]. \tag{4.31}$$

Thus, obtaining Eq. (4.31), we can eliminate the equation for the vibrational temperature perturbation from system (4.13)–(4.16) and reduce the order of the system to sixth. The reduced system has two trivial solutions: $u = v = \theta = \theta_v = 0$ and $u = 1$, $v = -iy$, $\theta = \theta_v = 0$. Instead of them, one considers the following inviscid solutions as solutions of the original system (4.1)–(4.6):

$$V_k(y) = \left[u_{ki}(y),\ v_{ki}(y),\ \theta_{ki}(y)\right]^*, \quad k = 1, 2, \tag{4.32}$$

(the asterisk means that a vector-column of solutions is considered). Now one can construct two linearly independent solutions for the transverse velocity perturbation. Assuming in Eq. (4.13) that $\theta = 0$ and $u \neq 0$ and applying integration, we obtain

$$v^{(1)}_{1,2\mathrm{v}}(y) = -i\int_\infty^\zeta \frac{dy}{dY}\,d\xi \int_\infty^\xi \left(\frac{dY}{d\zeta}\right)^2\left(\frac{dy}{dY}\right)^{3/2} A_1\,(i\nu)\,d\nu$$

$$= -i\left(\frac{dy}{dY}\right)^{5/2}\left(\frac{dY}{d\zeta}\right)^2 A_{1,2}\,(i\zeta,\ 2)\,. \tag{4.33}$$

Here, as in Eq. (4.26) above, slowly changing functions as compared to

$$H^{(1)}_{1/3}\left(2\sqrt{z^3}/3\right)$$

are taken out from the integrands to write the equations in terms of $A_1\,(i\nu,\ 2)$.

Similarly, for $u = 0$ and $\theta \neq 0$, integrating Eq. (4.13), we obtain

$$v^{(2)}_{1,2\mathrm{v}}(y) = \frac{i(y-c)}{(\gamma+\gamma_v)\,T_s}\left(\frac{dY_\theta}{d\zeta_\theta}\right)\sqrt{\left(\frac{dy}{dY_\theta}\right)^3}\left[\gamma\,A_{1,2}\,(i\zeta_\theta,\ 1)\right.$$

$$- \frac{\gamma_v}{\sqrt{2}} A_{1,2}\left(i\sqrt[3]{2}\,\zeta_\theta,\ 1\right)\Big]. \tag{4.34}$$

As a result, the linearly independent solutions of the simplified "viscous" system are written as

$$\begin{aligned} V_3(y) &= \left[u_{1\mathrm{v}}(y),\ v_{1\mathrm{v}}^{(1)}(y),\ 0\right]^*, \quad V_4(y) = \left[0,\ v_{1\mathrm{v}}^{(2)}(y),\ \theta_{1\mathrm{v}}\right]^*, \\ V_5(y) &= \left[u_{2\mathrm{v}}(y),\ v_{2\mathrm{v}}^{(1)}(y),\ 0\right]^*, \quad V_6(y) = \left[0,\ v_{2\mathrm{v}}^{(2)}(y),\ \theta_{2\mathrm{v}}\right]^*. \end{aligned} \tag{4.35}$$

4.2 Asymptotics of a Neutral Stability Curve of the Supersonic Couette Flow of a Vibrationally Excited Gas

Now we pass to constructing the neutral stability curve directly for the Couette flow described by Eqs. (4.17) and (4.18).

4.2.1 *Secular Equation and Its Solution*

To derive the spectral problem we have to require that the linear combination of the independent solutions (4.32), (4.35) should satisfy the boundary conditions (4.7). This is valid if

$$c_1 V_1(0) + c_2 V_2(0) + c_3 V_3(0) + c_4 V_4(0) + c_5 V_5(0) + c_6 V_6(0) = 0, \tag{4.36}$$

$$c_1 V_1(1) + c_2 V_2(1) + c_3 V_3(1) + c_4 V_4(1) + c_5 V_5(1) + c_6 V_6(1) = 0. \tag{4.37}$$

The homogeneous system (4.36), (4.37) has nontrivial solutions

$$(c_1,\ c_2,\ c_3,\ c_4,\ c_5,\ c_6)\ ,$$

when its determinant equals to zero:

$$\Delta = \begin{vmatrix} u_{1\mathrm{i}}(0) & u_{2\mathrm{i}}(0) & u_{1\mathrm{v}}(0) & 0 & u_{2\mathrm{v}}(0) & 0 \\ v_{1\mathrm{i}}(0) & v_{2\mathrm{i}}(0) & v_{1\mathrm{v}}^{(1)}(0) & v_{1\mathrm{v}}^{(2)}(0) & v_{2\mathrm{v}}^{(1)}(0) & v_{2\mathrm{v}}^{(2)}(0) \\ \theta_{1\mathrm{i}}(0) & \theta_{2\mathrm{i}}(0) & 0 & \theta_{1\mathrm{v}}(0) & 0 & \theta_{2\mathrm{v}}(0) \\ u_{1\mathrm{i}}(1) & u_{2\mathrm{i}}(1) & u_{1\mathrm{v}}(1) & 0 & u_{2\mathrm{v}}(1) & 0 \\ v_{1\mathrm{i}}(1) & v_{2\mathrm{i}}(1) & v_{1\mathrm{v}}^{(1)}(1) & v_{1\mathrm{v}}^{(2)}(1) & v_{2\mathrm{v}}^{(1)}(1) & v_{2\mathrm{v}}^{(2)}(1) \\ \theta_{1\mathrm{i}}(1) & \theta_{2\mathrm{i}}(1) & 0 & \theta_{1\mathrm{v}}(1) & 0 & \theta_{2\mathrm{v}}(1) \end{vmatrix} = 0\,. \tag{4.38}$$

For further simplification we estimate the order of magnitude of the elements of determinant (4.38). It follows from the expressions for the "inviscid" solutions in Sect. 4.1.1 that they are all bounded at $y = 0$ and $y = 1$, and one can assume that

$$u_{ki} \sim v_{ki} \sim \theta_{ki} \sim O(1) \quad k = 1, 2$$

on both boundaries at $\mathrm{M} \leq 4$.

In contrast to the bounded "inviscid" solutions, the "viscous" solutions (4.35) are expressed via the generalized Airy functions whose arguments on both boundaries are sufficiently large. Indeed, the following estimates are valid at the boundary points:

$$\zeta(y = 0) \sim \left[\frac{\alpha \mathrm{Re}}{T_s(0)}\right]^{1/3} c \sim O(10), \tag{4.39}$$

$$\zeta(y = 1) \sim \left[\frac{\alpha \mathrm{Re}}{T_s(y^*)}\right]^{1/3} (1 - c)^{3/2} \sim O\left(10^2\right), \tag{4.40}$$

These estimates are based on the characteristic values taken from the numerical calculations of the neutral curves described in Chap. 3:

$$\alpha \sim 2, \quad \mathrm{Re} \sim 10^6, \quad c \sim 10^{-2}.$$

This allows us to estimate the "viscous" elements in determinant (4.38) by using the universal asymptotic formula [2] for the generalized Airy functions:

$$A_{\pm}(x, p) \simeq \frac{(\pm 1)^p}{2\sqrt{\pi}} x^{-(2p+1)} \exp(\pm\eta), \quad \eta = \frac{2}{3} x^{3/2}. \tag{4.41}$$

In this case, we have $A_1(x, p) \sim A_-(x, p)$ and $A_2(x, p) \sim A_+(x, p)$.

It is seen from the above-given formula that the exponential factor plays a key role in the estimate. For convenience of the analysis we normalize the determinant columns corresponding to the "viscous" solutions $V_3, \ldots, V_6$ so that their maximum elements will have the order of $O(1)$, like the elements of the first two columns. For this purpose, in accordance with Eq. (4.41), one multiplies these columns in (4.38) by the normalization factors

$$N_3 \sim \zeta^3(0) \exp[\eta(0)], \quad N_4 \sim \zeta(0) \exp[\eta(0)],$$

$$N_5 \sim \zeta^3(1) \exp[-\eta(1)], \quad N_6 \sim \zeta(1) \exp[-\eta(1)].$$

As one can see in the columns V_3 and V_4, the elements calculated on the upper boundary at $y = 1$ are exponentially small. In columns V_5 and V_6, exponentially small elements are those calculated on the lower boundary at $y = 0$. Neglecting these elements, we present determinant (4.38) in the form

$$\Delta = \begin{vmatrix} u_{1i}(0) & u_{2i}(0) & \tilde{u}_{1v}(0) & 0 & 0 & 0 \\ v_{1i}(0) & v_{2i}(0) & \tilde{v}^{(1)}_{1v}(0) & \tilde{v}^{(2)}_{1v}(0) & 0 & 0 \\ \theta_{1i}(0) & \theta_{2i}(0) & 0 & \tilde{\theta}_{1v}(0) & 0 & 0 \\ u_{1i}(1) & u_{2i}(1) & 0 & 0 & \tilde{u}_{2v}(1) & 0 \\ v_{1i}(1) & v_{2i}(1) & 0 & 0 & \tilde{v}^{(1)}_{2v}(1) & \tilde{v}^{(2)}_{2v}(1) \\ \theta_{1i}(1) & \theta_{2i}(1) & 0 & 0 & 0 & \tilde{\theta}_{2v}(1) \end{vmatrix} = 0\,. \tag{4.42}$$

Here the tilde is used to mark the normalized nonzero elements.

Let us consider the thus-transformed subsystem (4.37) at $y = 1$. By virtue of homogeneity the unknowns involved in this system can be renormalized so to eliminate the unknown c_2. In the new variables the subsystem is written as

$$\begin{aligned} &C_1 u_{1i}(1) + C_5 \tilde{u}_{2v}(1) = -u_{2i}(1)\,, \\ &C_1 v_{1i}(1) + C_5 \tilde{v}^{(1)}_{2v}(1) + C_6 \tilde{v}^{(2)}_{2v}(1) = -v_{2i}(1)\,, \\ &C_1 \theta_{1i}(1) + C_6 \tilde{\theta}_{2v}(1) = -\theta_{2i}(1). \end{aligned} \tag{4.43}$$

Let us now consider the subsystem at $y = 0$, which is also presented in the renormalized form as

$$\begin{aligned} &C_1 u_{1i}(0) + C_3 \tilde{u}_{1v}(0) = -u_{2i}(0)\,, \\ &C_1 v_{1i}(0) + C_3 \tilde{v}^{(1)}_{1v}(0) + C_4 \tilde{v}^{(2)}_{1v}(0) = -v_{2i}(0)\,, \\ &C_1 \theta_{1i}(0) + C_4 \tilde{\theta}_{1v}(0) = -\theta_{2i}(0). \end{aligned} \tag{4.44}$$

To derive the secular equation it is sufficient to express C_1 from subsystem (4.43) because subsystem (4.44) does not include the unknowns C_5 and C_6. Obviously,

$$C_1 = \Delta_1(1)/\Delta(1),$$

where

$$\Delta(1) = \tilde{u}_{2v}(1)\tilde{\theta}_{2v}(1) \begin{vmatrix} u_{1i}(1) & 1 & 0 \\ v_{1i}(1) & \tilde{v}^{(1)}_{2v}/\tilde{u}_{2v}(1) & \tilde{v}^{(2)}_{2v}(1)/\tilde{\theta}_{2v}(1) \\ \theta_{1i}(1) & 0 & 1 \end{vmatrix}$$

$$\Delta_1(1) = \tilde{u}_{2v}(1)\tilde{\theta}_{2v}(1) \begin{vmatrix} -u_{2i}(1) & 1 & 0 \\ -v_{2i}(1) & \tilde{v}^{(1)}_{2v}/\tilde{u}_{2v}(1) & \tilde{v}^{(2)}_{2v}(1)/\tilde{\theta}_{2v}(1) \\ -\theta_{2i}(1) & 0 & 1 \end{vmatrix}$$

From here, it follows that

$$C_1 = \frac{v_{2i}(1) - u_{2i}(1)\tilde{v}^{(1)}_{2v}/\tilde{u}_{2v}(1) + \theta_{2i}(1)\tilde{v}^{(2)}_{2v}(1)/\tilde{\theta}_{2v}(1)}{-v_{1i}(1) + u_{1i}(1)\tilde{v}^{(1)}_{2v}/\tilde{u}_{2v}(1) + \theta_{1i}(1)\tilde{v}^{(2)}_{2v}(1)/\tilde{\theta}_{2v}(1)}. \tag{4.45}$$

However, after normalization of the columns to eliminate the exponentially small elements, the relations for the renormalized elements in Eq. (4.45) include the power-law multipliers. Indeed, it follows from Eq. (4.41) and from the expressions for the normalization coefficients that

$$\tilde{v}^{(1)}_{2\mathrm{v}}/\tilde{u}_{2\mathrm{v}}(1) \sim \zeta^{-2}, \quad \tilde{v}^{(2)}_{2\mathrm{v}}(1)/\tilde{\theta}_{2\mathrm{v}}(1) \sim \zeta^{-2}$$

Therefore, taking into account Eq. (4.40), we can write

$$C_1 = -\frac{v_{2\mathrm{i}}(1)}{v_{1\mathrm{i}}(1)} + O\left[(\alpha\mathrm{Re})^{-2/3}\right] \tag{4.46}$$

This means, in particular, that the required nonpenetration condition (4.7) for the transverse velocity perturbation is satisfied on the upper boundary $y = 1$ with accuracy to $O\left[(\alpha\mathrm{Re})^{-2/3}\right]$.

The first and third equations of subsystem (4.44) yield, respectively,

$$C_3 = -\frac{C_1 u_{1\mathrm{i}}(0) + u_{2\mathrm{i}}(0)}{\tilde{u}_{1\mathrm{v}}(0)}, \quad C_4 = -\frac{C_1 \theta_{1\mathrm{i}}(0) + \theta_{2\mathrm{i}}(0)}{\tilde{\theta}_{1\mathrm{v}}(0)}. \tag{4.47}$$

Substituting Eqs. (4.46) and (4.47) into the second equation of subsystem (4.44), we have the sought secular equation

$$C_1 v_{1\mathrm{i}}(0) + v_{2\mathrm{i}}(0) - \frac{C_1 u_{1\mathrm{i}}(0) + u_{2\mathrm{i}}(0)}{u_{1\mathrm{v}}(0)} v^{(1)}_{1\mathrm{v}}(0) - \frac{C_1 \theta_{1\mathrm{i}}(0) + \theta_{2\mathrm{i}}(0)}{\theta_{1\mathrm{v}}(0)} v^{(2)}_{1\mathrm{v}}(0) = 0\,.$$

Here we returned to nonnormalized "viscous" solutions.

For further transformations of the resultant equation we express the relations for the "viscous" solutions via the tabulated Tietjens functions [2, 4, 11]. In particular replacing the variable $Z = -\zeta$ in Eqs. (4.26) and (4.33), we receive

$$\frac{v^{(1)}_{1\mathrm{v}}(0)}{u_{1\mathrm{v}}(0)} = iZ\frac{dY}{d\zeta}\left(\frac{dy}{dY}\right)\left[-\frac{A_1(-Z,\,2)}{ZA_1(-Z,\,1)}\right] = iZ\frac{dY}{d\zeta}\left(\frac{dy}{dY}\right)F(Z)\,, \tag{4.48}$$

where $F(Z)$ is the Tietjens function for which detailed tables are available in [2, 4] (see also [12]). As in [4], one introduces the auxiliary function

$$\widetilde{G}(Z) = -\frac{A_1(-Z,\,2)}{ZA_1(-Z,\,1)}, \tag{4.49}$$

which can also be expressed via the Tietjens function and its derivative, and which is also tabulated in [4].

The ratio of the "viscous" solutions (4.34) and (4.31) can be presented as

$$\frac{v_{1v}^{(2)}(0)}{\theta_{1v}(0)} = -\frac{ic}{T_s}\left(\frac{dy}{dY_\theta}\right)\left(\frac{dY_\theta}{d\zeta_\theta}\right)J,$$

where

$$J = \frac{\gamma A_1(i\zeta_\theta,\ 1) - \gamma_v A_1\left(i\sqrt[3]{2}\,\zeta_\theta,\ 1\right)/\sqrt{2}}{\gamma A_1(i\zeta_\theta) - \gamma_v A_1\left(i\sqrt[3]{2}\,\zeta_\theta\right)/\sqrt[3]{2}}. \tag{4.50}$$

Expression (4.50) is transformed to be presented via the functions $F(Z)$ and $\widetilde{G}(Z)$:

$$J = \frac{\left(\gamma - \gamma_v/\sqrt{2}\right)A_1(i\zeta_\theta,\ 1) - \gamma_v \int\limits_{i\zeta_\theta}^{i\zeta_\theta/\sqrt[3]{2}} A_1(\zeta)\,d\zeta/\sqrt{2}}{\left(\gamma - \gamma_v/\sqrt[3]{2}\right)A_1(i\zeta_\theta) - \gamma_v\left[A_1\left(i\sqrt[3]{2}\,\zeta_\theta\right) - A_1(i\zeta_\theta)\right]/\sqrt[3]{2}}$$

Then the integral in the numerator is presented by the formula of left rectangles and the difference in the denominator is presented by the Lagrangian formula of finite accretions

$$J = \frac{\left(\gamma - \gamma_v/\sqrt{2}\right)A_1(i\zeta_\theta,\ 1)}{\left(\gamma - \gamma_v/\sqrt[3]{2}\right)A_1(i\zeta_\theta)}\left[\frac{1 - \gamma_v\left(\gamma\sqrt{2} - \gamma_v\right)^{-1}\zeta_\theta\left(\sqrt[3]{2} - 1\right)\dfrac{A_1(\zeta_\theta)}{A_1(\zeta_\theta,\ 1)}}{1 - \gamma_v\left(\gamma\sqrt[3]{2} - \gamma_v\right)^{-1}\zeta_\theta\left(\sqrt[3]{2} - 1\right)\dfrac{A_1'(\zeta_\theta)}{A_1(\zeta_\theta)}}\right]$$

The derivative of the Airy function of the first kind is expressed by the recurrent formula (4.24) at $p = 2$:

$$A_1(\zeta_\theta,\ -1) \equiv A_1'(\zeta_\theta) = \zeta_\theta A_1(\zeta_\theta,\ 1) - A_1(\zeta_\theta,\ 2).$$

The ratio of the generalized Airy functions in the denominator is presented as

$$\frac{\zeta_\theta A_1(\zeta_\theta,\ 1) - A_1(\zeta_\theta,\ 2)}{A_1(\zeta_\theta,\ 1)} = \frac{\zeta_\theta A_1(\zeta_\theta,\ 1)}{A_1(\zeta_\theta,\ 1)}\left(1 - \frac{A_1(\zeta_\theta,\ 2)}{\zeta_\theta A_1(\zeta_\theta,\ 1)}\right).$$

Substituting the resultant expression into the last formula for J and applying the replacement $Z_\theta = -\zeta_\theta$, we obtain the sought expression in terms of the functions $F(Z_\theta)$ and $\widetilde{G}(Z_\theta)$:

$$J = \frac{\left(\gamma\sqrt{2} - \gamma_v\right)Z_\theta}{\sqrt[6]{2}\left(\gamma\sqrt[3]{2} - \gamma_v\right)}$$

$$\times \left[\frac{\widetilde{G}(Z_\theta) - \gamma_v \left(\gamma\sqrt{2} - \gamma_v\right)^{-1} \left(\sqrt[3]{2} - 1\right) Z_\theta}{1 + \left(\sqrt[3]{2} - 1\right)\left(\gamma\sqrt[3]{2} - \gamma_v\right)^{-2} Z_\theta^3 \widetilde{G}(Z_\theta)\left(1 - F(Z_\theta)\right)} \right] \tag{4.51}$$

or $J = -Z_\theta J_1(Z_\theta)$ in the form convenient for further transformations. As a result, one obtains

$$\frac{v_{1v}^{(2)}(0)}{\theta_{1v}(0)} = \frac{ic}{T_s}\left(\frac{dy}{dY_\theta}\right)\left(\frac{dY_\theta}{d\zeta_\theta}\right) Z_\theta J_1(Z_\theta). \tag{4.52}$$

For a perfect gas without excitation of vibrational mode when $\gamma_v = \theta_v = 0$, the relaxation terms in the simplified system become equal to zero, and the system transforms to the Dunn-Lin "viscous" system.

As in [4], we have

$$\frac{v_{1v}^{(2)}(0)}{\theta_{1v}(0)} = \frac{ic}{T_s}\left(\frac{dy}{dY_\theta}\right)\left(\frac{dY_\theta}{d\zeta_\theta}\right) Z_\theta \widetilde{G}(Z_\theta). \tag{4.53}$$

Let us transform the factor in formula (4.48):

$$\widetilde{M} = Z\frac{dY}{d\zeta}\left(\frac{dy}{dY}\right)\bigg|_{y=0} = -(\alpha\mathrm{Re})^{1/3}\, Y \frac{1}{(\alpha\mathrm{Re})^{1/3}} Y^{1/2}\frac{\sqrt{T_s}}{\sqrt{-c}} = -Y^{3/2}\frac{\sqrt{T_s}}{\sqrt{-c}}.$$

In view of Eq. (4.19), we present

$$\widetilde{M} = \frac{3}{2}\frac{\sqrt{T_s}}{\sqrt{-c}}\int\limits_0^c \sqrt{\frac{t-c}{T_s(t)}}\, dt.$$

Redenoting the factor $\widetilde{M}$ similar to [4], we can write

$$\widetilde{M} = (1+\lambda)\frac{c}{U_s'},$$

where

$$1+\lambda = \frac{3}{2}\frac{\sqrt{T_s}}{c^{3/2}}\int\limits_0^c \sqrt{\frac{t-c}{T_s(t)}}\, dt,$$

and Eq. (4.48) is transformed to

$$\frac{v_{1v}^{(1)}(0)}{u_{1v}(0)} = i(1+\lambda)\, cF(Z). \tag{4.54}$$

Let us demonstrate that

$$Z\frac{dY}{d\zeta}\left(\frac{dy}{dY}\right)\Big|_{y=0} = Z_\theta\left(\frac{dy}{dY_\theta}\right)\frac{dY_\theta}{d\zeta_\theta}\Big|_{y=0}.$$

Indeed, as $Z_\theta = \mathrm{Pr}^{1/3}Z$ (see Eq. (4.29)), then

$$Z_\theta\left(\frac{dy}{dY_\theta}\right)\frac{dY_\theta}{d\zeta_\theta}\Big|_{y=0} = \mathrm{Pr}^{1/3}Z\frac{1}{\mathrm{Pr}^{1/3}}\frac{dy}{dY}\left(\frac{dY}{d\zeta}\right) = Z\frac{dY}{d\zeta}\left(\frac{dy}{dY}\right)\Big|_{y=0} \tag{4.55}$$

To transform the inviscid terms of the secular equation we introduce the notations

$$u_{\mathrm{i}}(0) = C_1 u_{1\mathrm{i}}(0) + u_{2\mathrm{i}}(0), \quad v_{\mathrm{i}}(0) = C_1 v_{1\mathrm{i}}(0) + v_{2\mathrm{i}}(0),$$
$$\theta_{\mathrm{i}}(0) = C_1\theta_{1\mathrm{i}}(0) + \theta_{2\mathrm{i}}(0). \tag{4.56}$$

The expressions for the functions of the inviscid perturbations of the streamwise and transverse velocities and static temperature from Sect. 4.1.1 yield the relation

$$\frac{\theta(0)}{u(0)} = (\gamma-1)\,\mathrm{M}^2\Delta(0)\left[c+\frac{iv(0)}{u(0)}\right], \quad \Delta(0) = 1 - \frac{\tau\gamma_v}{\gamma+\gamma_v-i\gamma\alpha\tau c}.$$

In view of the introduced relations and also Eqs. (4.52), (4.54), and (4.55), the secular equation can be written in the form

$$\frac{v_{\mathrm{i}}(0)}{u_{\mathrm{i}}(0)} = \frac{i(1+\lambda)\,c\left[F(Z) + \dfrac{(\gamma-1)\mathrm{M}^2c^2\Delta(0)}{T_s(0)}J_1\right]}{1+\dfrac{(\gamma-1)\mathrm{M}^2c\Delta(0)}{T_s(0)}(1+\lambda)\,cJ_1} \tag{4.57}$$

If the notation $P = C_1p_1 + p_2$ is used, the expressions for the inviscid solutions and Eq. (4.56) yield

$$u_{\mathrm{i}}(0) = \frac{T_s(0)}{\alpha c}\left(-\frac{P'}{\alpha c}+\alpha P\right), \quad v_{\mathrm{i}}(0) = \frac{T_s(0)}{i\alpha^2 c}P'.$$

Let

$$\Pi(0) = \frac{P'(0)}{\alpha^2 P(0)},$$

then the ratio in the left-hand side of Eq. (4.57) can be rewritten as

$$\frac{v_{\mathrm{i}}(0)}{u_{\mathrm{i}}(0)} = \frac{\Pi(0)}{i(1-\Pi(0)/c)},$$

whence it follows, in turn, that

$$\Pi(0) = \frac{c}{1 - ic\,[v_{\mathrm{i}}(0)/u_{\mathrm{i}}(0)]^{-1}}.$$

As a result, the secular equation (4.57) transforms to

$$\Pi(0) = -c\,\frac{F(Z) + KJ_1(Z_\theta)}{(1+\lambda)^{-1} - F(Z)}, \tag{4.58}$$

where

$$K = \frac{(\gamma - 1)\,\mathrm{M}^2 c^2 \Delta(0)}{T_s(0)}.$$

4.2.2 Asymptotics of the Critical Reynolds Numbers and Branches of the Neutral Stability Curve

In previous publications (see, e.g., [10, 11]) the functions $F(Z)$ and $\widetilde{G}(Z)$ are tabulated up to $Z \le 10$. However, our calculations require the secular equation to be solved up to greater values of Z. For this purpose, we had to derive the asymptotics of the right-hand-side of Eq. (4.58) as $Z \to \infty$.

In the limit of large values of the argument the asymptotics of the Tietjens function has the following form [1]:

$$F(Z) \approx \left(Z^{3/2} \exp\left(-\frac{\pi}{4}\right) - \frac{5}{4}\right)^{-1}. \tag{4.59}$$

Let us denote $q = Z^{-3/2}/\sqrt{2}$. In terms of the introduced variable, we have the following expression with accuracy to the terms $O\left(q^2\right)$:

$$F(Z) \approx q + iq\left(1 + \frac{5q}{2}\right)^{-1}.$$

For the auxiliary function $\widetilde{G}(Z_\theta)$ we derive the following expression [4]:

$$\widetilde{G}(Z_\theta) = \frac{F(Z_\theta)}{1 - F(Z_\theta) - Z_\theta F'(Z_\theta)}.$$

Differentiating Eq. (4.59) and applying simple transformations we receive the estimate

$$\widetilde{G}(Z_\theta) \approx \frac{1}{\sqrt{\mathrm{Pr}}}\left[q + iq\left(1 + \frac{3q}{2\sqrt{\mathrm{Pr}}}\right)\right].$$

In view of the asymptotic relations obtained above we find the sought form of the secular equation for large values of Z:

$$\Pi(0) = -c(1+\lambda)q\left\{1 + i + 2qi(1+\lambda) + q\left[\frac{5i}{2} + K\mathrm{Pr}^{-1/2}(1+i)\right]\right\}. \quad (4.60)$$

In the theory of neutral stability curve it is of significant interest to estimate the critical Reynolds number of the considered flow. Such an estimate was obtained for a compressible boundary layer on a flat plate in [1] (see also [11]), where an explicit expression was derived for Re_{cr} as a function of the free-stream parameters.

Similar to [1], let us consider the range of moderate supersonic Mach numbers $\mathrm{M} \leq 4$. In this case the right-hand side of Eq. (4.58) is simplified. Indeed, one has $1+\lambda \leq 1.1$, which allows to neglect of the value of λ as compared to unity in the first approximation. Moreover, it follows from our numerical calculations of the neutral stability curve in Chap. 3 that the characteristic value of the phase velocity of the neutral perturbations along this curve lies in the interval $c \approx 0.1$–0.2. This gives one an estimate

$$K = \frac{(\gamma - 1)\,\mathrm{M}^2 c^2 \Delta(0)}{T_s(0)} \approx 0.02\text{–}0.075.$$

As $F(Z) \sim J_1(Z_\theta)$, the second term in the numerator in Eq. (4.58) can be neglected. As a result, one can write the approximate secular equation

$$\Pi(0) = \frac{F(Z)}{1 - F(Z)} = \frac{1}{1 - F(Z)} - 1 = \mathscr{F}(Z) - 1, \quad (4.61)$$

where $\mathscr{F}(Z)$ is the modified Tietjens function whose tabulated value can be also found in many origins (see, e.g., [10]). Separating the real and imaginary parts in the resultant equation, we obtain the following system:

$$\Pi_r(0) = \mathscr{F}_r(Z) - 1, \quad \Pi_i(0) = \mathscr{F}_i(Z).$$

For calculating the critical Reynolds number, it should be noted that the function $\mathscr{F}_i(Z)$ is a bell-shaped curve and its maximum value $\mathscr{F}_i(Z) = 0.58$ corresponds to the value of the argument $Z = 3.24$. The latter corresponds to $\mathscr{F}_r(Z) = 1.46$. The branches of the function $\mathscr{F}_i(Z)$ determine the lower and upper branches of the neutral curve. The merging point of these branches in the plane (Re, α) corresponds to Re_{cr}. Therefore, the system for estimating the critical Reynolds number can be written as follows

$$\Pi_r(0) = 0.46, \quad \Pi_i(0) = 0.58. \quad (4.62)$$

System (4.62) allows one to calculate the unknown critical values of the phase velocity c_{cr} and the wave number α_{cr}. In system (4.62) it is impossible to pass to simpler relations, as it was done in [1], primarily because the wavenumbers on the neutral curve are not small and lie in the range $\alpha \sim 2.4$–2.7. The use of asymptotics only for small values of c_{cr} does not lead to significant simplification of the left-hand sides of Eqs. (4.62). After finding the solution of system (4.62) (c_{cr}, α_{cr}), one can calculate the approximate value of the critical Reynolds number by the formula

$$\mathrm{Re}_{cr} = \frac{3.24^3}{\alpha_{cr}\left[\dfrac{3}{2}\displaystyle\int_0^c \sqrt{\dfrac{c-y}{T_s(y)}}\,dy\right]^2} = \frac{3.24^3\, T_s(0)}{\alpha_{cr}\, c_{cr}^3}, \tag{4.63}$$

where the integral in the denominator for this temperature profile is approximately calculated with accuracy up to the fourth decimal place after point.

It is of certain interest to consider the far asymptotic behavior of the branches of the neutral stability curves, as done in the classical studies [2, 4]. Let us confine ourselves to the approximation considered above. The secular equation (4.61) is considered in the following form (cf. [6]):

$$(u+1) + iv = \frac{1}{1-F(Z)} = \mathscr{F}(Z) = \mathscr{F}_r(Z) + i\mathscr{F}_i(Z)\,. \tag{4.64}$$

As Z $\to \infty$, the following asymptotic presentation is valid:

$$\mathscr{F}(Z) = 1 + i\frac{1}{\sqrt{2Z^3}}\,. \tag{4.65}$$

As the argument Z decreases from its large values, $\mathscr{F}_i(Z)$ on the right branch increases to its maximum value $\mathscr{F}_i(3.24) = 0.58$. After that, $\mathscr{F}_i(Z)$ on the left branch monotonically decreases and vanishes at $Z = 2.294$. As was demonstrated above, the following relation is satisfied with sufficient accuracy:

$$\alpha\mathrm{Re} = Z^3\left[\frac{3}{2}\int_0^c \sqrt{\frac{c-y}{T_s(y)}}\,dy\right]^{-2} = \frac{Z^3 T_s(0)}{c^3}.$$

It follows that such a relation is valid on the lower branch of the neutral stability curve:

$$\mathrm{Re}_d = \frac{2.294^3}{\alpha c^3}\, T_s(0).$$

In turn, as $Z \to \infty$, it follows from Eqs. (4.64) and (4.65) that $u = 0$ and

$$v = 1/\sqrt{2Z^3},$$

whence there follows the asymptotic expression for the upper branch:

$$\alpha \mathrm{Re} = \frac{T_s(0)}{2vc^3}.$$

The asymptotic estimate of the imaginary "inviscid" part of the secular equation (4.62) as $c \to 0$ yields the approximate dependence $v \sim \pi A c \sim c$, and the expression for the upper branch is

$$\mathrm{Re}_u \sim \frac{T_s(0)}{2\alpha c^5}.$$

To obtain far asymptotics of branches on the plane (α, Re), we have to know the dependence $c \approx \alpha^k$ as $c \to 0$. However, as the numerical calculations show, the wavenumber here is not very small with variation of the Reynolds number within three orders of magnitude $\mathrm{Re} \in \left(10^5-10^8\right)$ and has the order approximately $O(1)$, This is in contrast to the available results for the boundary layer [1, 4] and incompressible Poiseuille flow [2], Therefore, it is impossible to obtain an explicit estimate from the analysis of the "viscous" part of the secular equation. It follows from the numerically calculated neutral curves (see Chap. 3) that $c \approx 2 \times 10^{-3}\alpha^4$. As a result, we can obtain only intermediate asymptotic solutions in the range of Reynolds numbers $\mathrm{Re} \in \left(10^5-10^8\right)$ in the following form.

For the lower branch of the neutral curve a searched asymptotics is

$$\mathrm{Re}_d \sim \frac{2.294^3}{8\alpha^{13}} 10^9 \, T_s(0).$$

For the upper branch the correspondent relation looks as follows

$$\mathrm{Re}_u \sim \frac{T_s(0)}{256\,\alpha^{21}} 10^{15}.$$

4.2.3 Numerical Calculations of Secular Equation

The secular equation (4.58) has a typical structure similar to equations considered, e.g., in [4, 6]. Its left-hand "inviscid" part depends on the phase velocity c and wavenumber α. At the same time the right "viscous" part is expressed via the tabulated functions of the variable Z and also depends on c. Therefore, the points on the neutral stability curves Re (α, γ_v, M) on the plane (Re, α) were calculated in the same sequence as those in [4, 6].

For fixed values of the flow parameters, the Mach number M, and the degree of vibrational excitation γ_v, the values of the phase velocity c were set in the interval $c = [0;\ 1]$ with a step $\Delta c = 10^{-3}$. The integral in the right-hand side of Eq. (4.58) was calculated using the Simpson's formula [12]. The arrays of real and imaginary values of the right-hand side as functions of Z were calculated for $Z = [0;\ 10]$ with the use of the tables [4]. For $Z = [10;\ 50]$, the asymptotic formula (4.60) was used. In both cases, the table step $\Delta Z = 0.04$ was chosen [4].

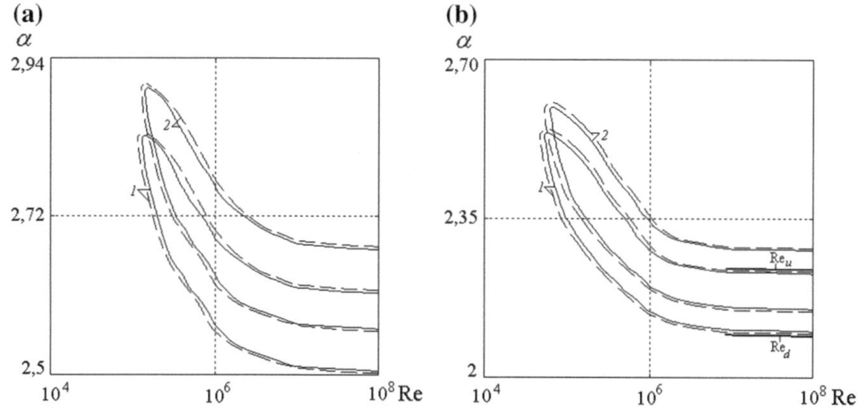

Fig. 4.1 Neutral stability curves Re(α) for modes I and II for M $= 3$, $\gamma_v = 0$ (*1*) and $\gamma_v = 0.667$, $\tau = 1$ (*2*). **a** is mode I. **b** is mode II

After calculating the arrays of the right-hand side the arrays of the left-hand side of Eq. (4.58) for a given value of c were calculated as a function of the wavenumber α in the interval $\alpha = [1;\ 3]$ with a step $\Delta\alpha = 10^{-3}$. The calculated arrays for the right-hand and left-hand sides were compared until identical values within 10^{-8} were reached if it was possible for the fixed value of c. After that, the calculation was repeated for the next value of the phase velocity. As a result, the arrays of the wavenumbers $\overline{\alpha}_k$, phase velocities $\overline{c}_k$, and variable $\overline{Z}_k$ corresponding to the points on the neutral curve were obtained. Using the formula

$$\mathrm{Re}_k(\overline{\alpha}_k) = \frac{\overline{Z}_k^3}{\overline{\alpha}_k}\left[\frac{3}{2}\int\limits_0^{\overline{c}_k}\sqrt{\frac{\overline{c}_k - y}{T_s(y)}}\,dy\right]^{-2},$$

which follows from the replacement of the independent variable in Eq. (4.21), we calculated the values of the Reynolds number Re_k on the neutral curve. The neutral curves for modes I and II are identified on the basis of the intervals of the phase velocity. At $\overline{c}_k \in (0.5;\ 1)$ the points belong to the neutral stability curve of mode I; otherwise, at $\overline{c}_k \in (0;\ 0.5)$, the points correspond to the neutral stability curve of mode II.

Figures 4.1 and 4.2 show the calculated neutral stability curves in semilogarithmic coordinates for perfect and vibrationally excited gases for M $= 3$ and 4, respectively. The dotted curves are those predicted by solving the secular equation (4.58) of the asymptotic theory. For comparison the solid curves show the results of the direct numerical calculation of the original spectral problem (4.1)–(4.7). Because the asymptotic values are shown on the lower Re_d and upper Re_u neutral curves in the domain Re $\sim 10^8$. For comparatively small Mach numbers, the asymptotic

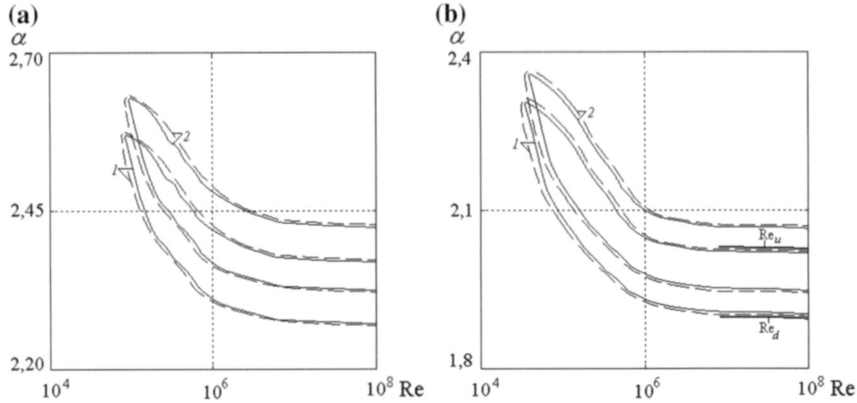

Fig. 4.2 Neutral stability curves $Re(\alpha)$ for modes I and II for $M = 4$, $\gamma_v = 0$ (*1*) and $\gamma_v = 0.667$, $\tau = 1$ (*2*). **a** is mode I. **b** is mode II

Table 4.1 Critical Reynolds numbers Re_{cr} and wavenumbers α_{cr} for modes I and II (numerical calculation of the complete spectral problem (4.1)–(4.7))

γ_v	$M = 3$		$M = 4$	
	Re_{cr}	α_{cr}	Re_{cr}	α_{cr}
Mode I				
0	123900	2.835	82703	2.570
0.667	137303	2.901	91650	2.628
Mode II				
0	50060	2.546	33415	2.307
0.667	56061	2.603	37421	2.359

curves are in reasonable agreement with the results obtained for the original spectral problem. In particular the dissipative effect of vibrational relaxation is clearly visible.

It can be noted that the transition to the asymptotic theory slightly extends the domain of instability and reduces the critical Reynolds numbers. It is also seen that the limiting values of Re_d and Re_u agree well with the numerical predictions.

Tables 4.1 and 4.2 allow one to compare the critical Reynolds numbers and wavenumbers obtained by the numerical calculations of the original spectral problem and by the asymptotic theory. It is seen that the values of Re^{I}_{cr} and Re^{II}_{cr} for both modes predicted by the asymptotic theory are approximately 9–10% smaller than the corresponding values obtained in the numerical solution of the spectral problem (4.1)–(4.7). The deviation of the value of Re_{cr} obtained by solving the simplified system (4.62) lies in the same limits, which can be considered as a good approximation, at least for the perfect gas.

To conclude it should be noted that investigations of the plane Couette flow considered in this chapter is of interest for several reasons. First, an independent analytical confirmation is provided for the numerical results reported in Chap. 3. In addition it

Table 4.2 Critical Reynolds numbers Re_{cr} and wavenumbers α_{cr} for modes I and II (calculations by the asymptotic theory)

γ_v	M = 3		M = 4	
	Re_{cr}	α_{cr}	Re_{cr}	α_{cr}
Mode I				
0	112000	2.842	74800	2.576
0.667	124100	2.910	82800	2.639
Mode II				
0	44700	2.552	29700	2.314
0.667	50100	2.610	33400	2.365
Calculation by Eq. (4.63)				
0	41200	2.597	27400	2.354

is demonstrated that the use of the traditional asymptotic analysis of the influence of dissipative processes on flow stability yields correct results, in contrast to a simple "viscous" correction [13] of the phase velocities of the inviscid modes with the use of the boundary conditions. Finally, the spectral problem for the plane Couette flow with two boundary conditions has not been studied within the framework of the asymptotic theory even for the perfect gas.

References

1. Lin, C.C.: The Theory of Hydrodynamic Stability. University Press, New York (1966)
2. Drazin, P.G., Reid, W.H.: Hydrodynamic Stability. Cambridge University Press, Cambridge (2004)
3. Gordiets, B.F., Osipov, A.I., Shelepin, L.A.: Kinetic Processes in Gases and Molecular Lasers. Gordon and Breach, New York (1987)
4. Reshotko, E.: Stability of the Compressible Laminar Boundary Layer, Ph.D. thesis. California Institute of Technology, Pasadena (1960)
5. Naimark, M.N.: Linear Differential Operators. Nauka, Moscow (1969) (in Russian)
6. Dunn, D.W., Lin, C.C.: On the stability of the laminar boundary layer in a compressible fluid. J. Aero. Sci. **22**, 455–477 (1955)
7. Tamarkin, Ya.D.: Some General Problems of the Theory of Ordinary Differential Equations. Tipografiya M.P. Frolovoi, Petrograd (1917) (in Russian)
8. Fok, V.A.: Problems of Diffraction and Propagation of Radiowaves. Sovetskoye Radio, Moscow (1970) (in Russian)
9. Grigor'ev, Yu.N., Ershov, I.V.: Linear stability of the Couette flow of a vibrationally excited gas. 1. Inviscid problem. J. Appl. Mech. Tech. Phys. **55**, 258–269 (2014)
10. Miles, J.W.: The hydrodynamic stability of a thin film of liquid in uniform shearing motion. J. Fluid Mech. **8**, 593–610 (1960)
11. Gaponov, S.A., Maslov, A.A.: Development of Perturbations in Compressible Flows. Nauka, Novosibirsk (1980) (in Russian)
12. Korn, G.A., Korn, T.M.: Mathematical Handbook for Scientists and Engineers. McGraw-Hill, New York (1961)
13. Duck, P.W., Erlebacher, G., Hussaini, M.Y.: On the linear stability of compressible plane Couette flow. J. Fluid Mech. **258**, 131–165 (1994)

Chapter 5
Energy Theory of Nonlinear Stability of Plane Shear Flows of Thermally Nonequilibrium Gas

Abstract The energy stability theory extended by the authors to the case of compressible flows of a vibrationally excited molecular gas is used to study stability of a subsonic Couette flow. Universal approach is developed for derivation of equations of the energy balance of disturbances for energy functionals. Based on these equations variational problems are posed for determining the critical Reynolds number of the possible beginning of the laminar-turbulent transition. Their asymptotic solutions are obtained in the limit of long-wave disturbances and yield an explicit dependence of the critical Reynolds number on the bulk viscosity coefficient, Mach number, and vibrational relaxation time. Neutral stability curves are calculated for arbitrary wavenumbers on the basis of the numerical solution of eigenvalue problems. It is shown that the minimum critical Reynolds numbers in realistic (for diatomic gases) ranges of flow parameters increase with increasing bulk viscosity coefficient, Mach number, vibrational relaxation time, and degree of excitation of vibrational modes. The results obtained in the study qualitatively confirm the asymptotic estimates for critical Reynolds number.

This chapter deals with application of energy stability theory for compressible and vibrationally excited Couette flows of diatomic molecular gas. These studies were necessitated to theoretically consider the experimental results [1]. In the paper comparative experiments on laminar-turbulent transition in Hagen–Poiseuille flow in a round tube for nitrogen N_2 and carbon monoxide CO were performed. The thermodynamic and transport properties of these gases are almost identical but the bulk viscosity of CO calculated from data on ultrasound attenuation is several times higher that the similarly calculated value for N_2. It was found in the experiments that, under the equal conditions, the transition Reynolds number Re_t in the more "viscous" gas CO was approximately 10% higher than the corresponding value for N_2.

For some reasons, the validity of these results was questionable. In particular, for the bulk viscosities of the gases used there are different data (see Chap. 1 and references in [2]) obtained by measuring relaxation times in shock waves. From these data, which are also given in part in [1], it follows that the difference between the bulk viscosities N_2 and CO is small so that it cannot be responsible for the observed change in Re_t. The fact that in [1] there are no comments on this inconsistency was noted in [3].

Y.N. Grigoryev and I.V. Ershov, *Stability and Suppression of Turbulence in Relaxing Molecular Gas Flows*, Fluid Mechanics and Its Applications 117,
DOI 10.1007/978-3-319-55360-3_5

To estimate the effect of bulk viscosity on the nonlinear development of perturbations in experiments [1] it was needed to choose adequate mathematical model and tools. With this point of view the full Navier–Stokes equations for a viscous heat-conducting gas were considered in application to the compressible Couette flow which has analytical Navier–Stokes solutions, one of them was used in Chaps. 3 and 4. The experiments [1] were carried out at very small Mach numbers when Hagen–Poiseuille flow is practically incompressible one, which is absolutely stable with viewpoint of the linear stability theory [4]. The same refers to the plane Couette flow. Therefore in this case use of the linear theory it was pointless.

The only alternative was using an energy stability theory. It is commonly believed that such an approach would offers a real opportunity to take into account the nonlinear stage of instability, though in a generalized form [5]. But, as was known, the energy stability theory had not been previously applied for compressible flows because of an essential nonlinearity of the Navier–Stokes equations for a compressible heat-conducting gas. It is obvious that this factor is also important for two-temperature aerodynamics equations. Moreover, the critical Reynolds numbers predicted by the theory were in fact low-limit estimates, which were often well below the experimental data. However, using such considerations, we expected to obtain a difference between a thermal nonequilibrium gas and a perfect gas.

In Sect. 5.1 of the chapter, the energy balance equation for disturbances is derived within the framework of the Navier–Stokes model for a moderate level of thermal excitation. This equation allows one to formulate a variational problem for calculating the critical Reynolds number Re_{cr}, which defines the low limit of possible beginning of the laminar-turbulent transition. An asymptotic solution of the problem in the limit of long-wave disturbances is obtained, which provides an explicit dependence of Re_{cr} on bulk viscosity. Numerical solutions in a wide range of wavenumbers are obtained for a compressible Couette flow with a linear velocity profile. The results show that the critical Reynolds number increases with the growth of bulk viscosity within the range of realistic values for diatomic gases. This increase reaches approximately 30% as compared to the zero bulk viscosity case.

In Sect. 5.2 the same results for the Couette flow of a vibrationally excited gas described by a system of two-temperature aerodynamics equations are presented. In this case, the technique of derivation of the energy equation and the total energy functional are of universal interest. In particular, this approach may be useful in non-modal stability theory [6]. Asymptotic formulas for Re_{cr} are derived, which describe its dependence on the Mach number, bulk viscosity coefficient, and relaxation time. For arbitrary wavenumbers, the variational problem is numerically studied by the collocation method. It was demonstrated within the limits of the real regime parameters (M, γ_v, and τ) typical for diatomic gases that the minimum values of Re_{cr} are obtained in the case of longitudinal modes, and then Re_{cr} increases as these parameters grow. More than a twofold increase can be reached as compared to the equilibrium gas.

5.1 Energy Stability Analysis of a Plane Compressible Flow. Effect of a Bulk Viscosity

It is known the bulk viscosity in the Navier–Stokes equations takes into account the relaxation of internal molecular modes during moderate thermal excitation [7]. In this section the stability of the compressible Couette flow with a linear velocity profile is studied using energy theory. Previous known results of this theory on the stability of incompressible and inhomogeneous flows have been obtained taking into account the solenoidal nature of the admissible velocity fields [5], which is absent in compressible flows. Here difficulties of mathematical nature that arise in the last case have been successfully overcome. Some simplifications make it possible to completely solve the corresponding variational problem for this flow and obtain an explicit dependence of Re_{cr} on bulk viscosity.

5.1.1 Basic Equations and Functionals

The Couette flow stability problem is considered using the Navier–Stokes equations for a compressible viscous heat-conducting gas. The computational domain Ω is a rectangular parallelepiped, whose faces are parallel to the coordinate planes of the Cartesian system (x_1, x_2, x_3) and whose center coincides with the coordinate origin. The impenetrable infinite plates along which the main current is directed are perpendicular to the x_2 axis and are moved uniformly with velocities $\pm U_0$ in opposite directions.

The characteristic nondimensionalizing scales are the channel width L on the x_2 axis, the modulus of boundary plate velocity U_0, the density ρ_0 and temperature T_0 which is constant along any cross section of the channel, the time $\tau_0 = L/U_0$, and the pressure $p_0 = \rho_0 U_0^2$. In the dimensionless variables, the system of equations is written as

$$\frac{d\rho}{dt} + \rho\,\frac{\partial u_i}{\partial x_i} = 0,$$

$$\rho\,\frac{du_i}{dt} = -\frac{\partial p}{\partial x_i} + \frac{1}{\mathrm{Re}}\,\frac{\partial^2 u_i}{\partial x_j^2} + \frac{1}{\mathrm{Re}}\left(\alpha_1 + \frac{1}{3}\right)\frac{\partial^2 u_j}{\partial x_i\,\partial x_j},$$

$$\rho\frac{dT}{dt} + \gamma(\gamma-1)\mathrm{M}^2 p\,\frac{\partial u_i}{\partial x_i} = \frac{\gamma}{\mathrm{RePr}}\,\frac{\partial^2 T}{\partial x_i^2},$$

$$\gamma \mathrm{M}^2 p = \rho T, \quad \frac{d}{dt} = \frac{\partial}{\partial t} + u_i\,\frac{\partial}{\partial x_i}, \quad i, j = 1, 2, 3. \tag{5.1}$$

Here ρ, u_i, T, and p are the density, velocity components, temperature, and gas pressure, respectively; the summation is performed over repeated subscripts. It is assumed that the thermal capacity and dissipation coefficients in system (5.1) do not depend on temperature and are constant. The parameters included in Eqs. (5.1) are defined as follows: the coefficient α_1 is equal to the ratio of the bulk viscosity to the shear viscosity ($\alpha_1 = \eta_b/\eta$) and characterizes the degree of nonequilibrium of the internal degrees of freedom of the gas molecules; $\mathrm{M} = U_0/\sqrt{\gamma R T_0}$ is the Mach number of the main flow, $\mathrm{Re} = U_0 L \rho_0/\eta$ is the Reynolds number, $\mathrm{Pr} = \eta c_p/\lambda_0$ is the Prandtl number, R is the gas constant, $\gamma = c_p/c_v$ is the isentropic exponent, c_p are c_v are the specific heats at constant pressure and volume, respectively and λ_0 is the thermal conductivity. In the energy equation, the group of nonlinear terms constituting the so-called dissipation function are omitted. This approximation is widely used in stability problems for compressible flows [8, 9].

Plane Couette flow with a linear velocity profile, which is an exact steady-state solution of system (5.1), is described by the relations

$$\boldsymbol{U}_s(x_2) = (x_2, 0, 0), \quad T_s(x_2) = \rho_s(x_2) = 1, \quad p_s(x_2) = 1/\left(\gamma\,\mathrm{M}^2\right).$$

Representing the instantaneous values of the hydrodynamic quantities of the perturbed flow as

$$\rho = 1 + \rho', \quad u_i = U_{s,i} + u_i', \quad T = 1 + T', \quad p = 1/\left(\gamma\,\mathrm{M}^2\right) + p', \tag{5.2}$$

we write the equations for the disturbances ρ', u_i', T', and p' of the main flow without any constraint on their amplitudes:

$$\frac{\partial \rho'}{\partial t} + u_i \frac{\partial \rho'}{\partial x_i} + \rho \frac{\partial u_i'}{\partial x_i} = 0, \tag{5.3}$$

$$\rho\left(\frac{\partial u_i'}{\partial t} + u_j' \frac{\partial u_i'}{\partial x_j} + U_{s,j} \frac{\partial u_i'}{\partial x_j} + u_j' \frac{\partial U_{s,i}}{\partial x_j}\right)$$

$$= -\frac{\partial p'}{\partial x_i} + \frac{1}{\mathrm{Re}} \frac{\partial^2 u_i'}{\partial x_j^2} + \frac{1}{\mathrm{Re}} \left(\alpha_1 + \frac{1}{3}\right) \frac{\partial^2 u_j'}{\partial x_i\, \partial x_j}, \tag{5.4}$$

$$\rho\left(\frac{\partial T'}{\partial t} + u_j' \frac{\partial T'}{\partial x_j} + U_{s,j} \frac{\partial T'}{\partial x_j}\right) + \gamma(\gamma - 1)\mathrm{M}^2 p \frac{\partial u_i'}{\partial x_i} = \frac{\gamma}{\mathrm{RePr}} \frac{\partial^2 T'}{\partial x_i^2}, \tag{5.5}$$

$$\gamma\,\mathrm{M}^2 p' = \rho T' + \rho', \quad i, j = 1, 2, 3. \tag{5.6}$$

Equations (5.3)–(5.5) do not contain an explicit dependence of the unperturbed flow velocity (5.2) on the x_2 coordinate lest the form of summation over subscripts be complicated. It is assumed that, for $x_1 = \pm x_0/2$ and $x_3 = \pm z_0/2$, the disturbances of the velocity u_i', density ρ', and pressure p' satisfy the periodic boundary conditions, and on impenetrable boundaries $x_2 = \pm 1/2$, they vanish. For the temperature disturbance T', the following boundary conditions are specified:

$$\left.\frac{\partial T'}{\partial x_1}\right|_{x_1=-x_0/2} = \left.\frac{\partial T'}{\partial x_1}\right|_{x_1=+x_0/2}, \quad \left.\frac{\partial T'}{\partial x_2}\right|_{x_2=-1/2} = \left.\frac{\partial T'}{\partial x_2}\right|_{x_2=+1/2} = 0,$$

$$\left.\frac{\partial T'}{\partial x_3}\right|_{x_3=-z_0/2} = \left.\frac{\partial T'}{\partial x_3}\right|_{x_3=+z_0/2}.$$

Below, the dimensions of the domain Ω on the periodic (homogeneous) coordinates x_1 and x_3 are equal to the disturbance wavelength along the corresponding coordinate:

$$x_0 = \pi/\alpha, \quad z_0 = \pi/\delta.$$

Here α and δ are the moduli of the projections of the disturbance wave vector $\boldsymbol{k}$ on the x_1 and x_3 axes, respectively.

We define the kinetic energy of disturbances as an integral over the flow region in the form

$$E(t) = \int\limits_{\Omega} \frac{\rho u_i'^2}{2} \, d\Omega. \tag{5.7}$$

For the evolution of the quantity $E(t)$, from Eqs. (5.3) and (5.4), we derive the energy balance equation similarly to [10]. Equations (5.3) and (5.4) are multiplied by $u_i'^2$ and u_i', respectively, and are combined. On the left side of the resulting relation there is a series of terms in divergent form:

$$\frac{1}{2}\frac{\partial}{\partial t}\left(\rho u_i'^2\right) + \frac{1}{2}\frac{\partial}{\partial x_j}\left(\rho u_i'^2 u_j'\right) + \frac{1}{2}\frac{\partial}{\partial x_j}\left(\rho u_i'^2\right) + \rho u_i' u_j' \frac{\partial U_{s,i}}{\partial x_j}$$

$$= -u_i' \frac{\partial p'}{\partial x_i} + \frac{1}{\mathrm{Re}} u_i' \frac{\partial^2 u_i'}{\partial x_j^2} + \frac{1}{\mathrm{Re}}\left(\alpha_1 + \frac{1}{3}\right) u_i' \frac{\partial}{\partial x_i}\frac{\partial u_j'}{\partial x_j}. \tag{5.8}$$

Integration of equality (5.8) over the domain Ω transforms the divergent terms on the left side to integrals over the boundary, which vanish by virtue of the boundary conditions on the disturbances. The terms on the right side are integrated by parts, and the resulting boundary integrals also vanish. As a result, one obtains the integral equation

$$\frac{dE}{dt} \equiv \frac{d}{dt}\int\limits_{\Omega} \frac{\rho u_i'^2}{2} \, d\Omega = J_1 + J_2 - \frac{1}{\mathrm{Re}}\left(J_3 + \alpha_1 J_4\right). \tag{5.9}$$

The term

$$J_1 = -\int\limits_{\Omega} \rho u_i' u_j' \frac{\partial U_i}{\partial x_j} \, d\Omega \tag{5.10}$$

describes the energy exchange between the disturbances and the main flow. The integral

$$J_2 = \int\limits_{\Omega} p' \frac{\partial u_i'}{\partial x_i} d\Omega \tag{5.11}$$

can be treated as the work in pulsation compression (expansion) of the gas, and the integrals

$$J_3 = \int\limits_{\Omega} \left[\left(\frac{\partial u_i'}{\partial x_j} \right)^2 + \frac{1}{3} \left(\frac{\partial u_i'}{\partial x_i} \right)^2 \right] d\Omega, \quad J_4 = \int\limits_{\Omega} \left(\frac{\partial u_i'}{\partial x_i} \right)^2 d\Omega \tag{5.12}$$

correspond to energy dissipation.

In the above expressions the signs of the integrals J_1 and J_2 are not determined, whereas J_3 and J_4 are nonnegative. As the Reynolds number Re decreases to a certain value Re_{cr}, the dissipation terms J_3 and J_4 begin to dominate, the derivative $dE/dt < 0$ and any disturbances damp with time. This allows one to formulate a variational problem based on Eq. (5.9) to estimate the critical Reynolds number Re_{cr}, which corresponds to the condition $dE/dt = 0$ and is calculated as the minimum of the functional:

$$\mathrm{Re}_{cr} = \min \left(\frac{J_3 + \alpha_1 J_4}{J_1 + J_2} \right). \tag{5.13}$$

From equality (5.13), it follows that an increase in the bulk viscosity (or the parameter α_1) leads to an increase in the critical Reynolds number Re_{cr}, but to obtain a particular value of Re_{cr}, it is necessary to solve the variational eigenvalue problem [5].

Equation (5.9) was derived similarly to the equation for an incompressible fluid [5] and, in this form, it does not explicitly take into account the perturbation features in compressible flows. In particular, unlike for an incompressible fluid, the total disturbance energy in gases, especially in molecular gases, should contain not only the kinetic component $E(t)$ but also the internal energy in some form. In addition Eq. (5.9) does not contain an explicit dependence on the Mach number M. This is due to the fact that Eq. (5.9) was derived without using the energy equation (5.5) and the state equation (5.6).

The energy balance equation (5.9) can be transformed as follows. Using equality (5.2), the continuity equation (5.3), and the equation of state (5.6), one writes Eq. (5.5) as

$$p' \frac{\partial u_i'}{\partial x_i} = -\frac{\partial}{\partial t} \left(\frac{\rho T}{\gamma(\gamma - 1)\mathrm{M}^2} \right)$$

$$- \frac{1}{(\gamma - 1)\mathrm{M}^2} \frac{\partial}{\partial x_i} \left(u_i' + \mathrm{M}^2 u_i p' - \frac{1}{\mathrm{RePr}} \frac{\partial T'}{\partial x_i} \right). \tag{5.14}$$

After substitution of expression (5.14) into the integral J_2, the divergent terms vanish by virtue of the boundary conditions on the disturbances and on the left of Eq. (5.9) one has the time derivative of the integral [11]:

$$E_t(t) = \int\limits_{\Omega} \rho \left(\frac{u_i'^2}{2} + \frac{T}{\gamma(\gamma-1)\mathrm{M}^2} \right) d\Omega.$$

In view of the chosen nondimensionalization, it is easy to show that in the dimensional variables the term $\rho T / \left[\gamma(\gamma-1)\mathrm{M}^2\right]$ is the internal energy of a unit volume of the gas. Obviously, the energy functional E_t is positive definite. The transformed energy balance equation becomes

$$\frac{dE_t}{dt} = \Phi$$

$$\equiv -\int\limits_{\Omega} \left\{ (1+\rho')\, u_i' u_j' \frac{\partial U_{s,i}}{\partial x_j} + \frac{1}{\mathrm{Re}} \left[\left(\frac{\partial u_i'}{\partial x_j} \right)^2 + \left(\alpha_1 + \frac{1}{3} \right) \left(\frac{\partial u_i'}{\partial x_i} \right)^2 \right] \right\} d\Omega. \tag{5.15}$$

For (5.15) it is also possible to formulate a variational eigenvalue problem to find the critical Reynolds number Re_{cr}. To further simplify Eq. (5.15) we perform partial separation of the variables and write the dependencies of the disturbances of the velocity, density, and temperature on the periodic coordinate x_3 in the form

$$u_1' = u_1''(x_1,\, x_2) \cos(\delta x_3)\,, \quad u_2' = u_2''(x_1,\, x_2) \cos(\delta x_3)\,,$$

$$u_3' = u_3''(x_1,\, x_2) \sin(\delta x_3)\,, \quad \rho' = \rho''(x_1,\, x_2) \cos(\delta x_3)\,,$$

$$T' = T''(x_1,\, x_2) \cos(\delta x_3)\,. \tag{5.16}$$

At $x_1 = \pm\pi/\alpha$ the amplitude functions u_i'', ρ'', and T'' satisfy the periodic boundary conditions, and on the impenetrable boundaries $x_2 = \pm 1/2$, they vanish. Using representation (5.16) in Eq. (5.15) we perform integration over the variable x_3 in the range $[-\pi/\delta;\, \pi/\delta]$. As shown in [12] the operations of variation and partial integration over homogeneous coordinates are permutable and a change in their order does not change the original variational problem. As a result, we have

$$\frac{dE_t''}{dt} = \Phi'' \equiv -\int\limits_{S} \left\{ u_1'' u_2'' + \frac{1}{\mathrm{Re}} \left[\left(\frac{\partial u_1''}{\partial x_1} \right)^2 + \left(\frac{\partial u_1''}{\partial x_2} \right)^2 + \left(\frac{\partial u_2''}{\partial x_1} \right)^2 \right.\right.$$

$$+ \left(\frac{\partial u_2''}{\partial x_2} \right)^2 + \left(\frac{\partial u_3''}{\partial x_1} \right)^2 + \left(\frac{\partial u_3''}{\partial x_2} \right)^2 + \delta^2 \left(u_1''^2 + u_2''^2 + u_3''^2 \right)$$

$$\left.\left. + \left(\alpha_1 + \frac{1}{3} \right) \left(\frac{\partial u_1''}{\partial x_1} + \frac{\partial u_2''}{\partial x_2} + \delta u_3'' \right)^2 \right] \right\} dS. \tag{5.17}$$

From expression (5.17) it follows that, after transformation (5.15), the varied functional Φ'' on the right side becomes quadratic in the amplitude functions u_i''. Naturally, the transformation (5.16) restricts a class of admissible functions, but as a result one derives a linear differential eigenvalue problem. This allows us to obtain asymptotical solutions in explicit form and simplifies numerical calculations.

5.1.2 *Variational Problem. Quality Properties and Asymptotics of Low Critical Reynolds Numbers*

Subjecting the functions u_k'' in the functional Φ'' to small smooth variations $u_k'' + \delta u_k''$ admitted by the boundary conditions, one distinguishes a functional $L\left(\delta u_k''\right)$, which is linear in the variation vector and leads to the Euler-Lagrange equations

$$\Delta_2 u_1'' + \left(\alpha_1 + \frac{1}{3}\right)\frac{\partial}{\partial x_1}\left(\frac{\partial u_1''}{\partial x_1} + \frac{\partial u_2''}{\partial x_2} + \delta u_3''\right) = \frac{\mathrm{Re}}{2}u_2'',$$

$$\Delta_2 u_2'' + \left(\alpha_1 + \frac{1}{3}\right)\frac{\partial}{\partial x_2}\left(\frac{\partial u_1''}{\partial x_1} + \frac{\partial u_2''}{\partial x_2} + \delta u_3''\right) = \frac{\mathrm{Re}}{2}u_1'',$$

$$\Delta_2 u_3'' - \delta\left(\alpha_1 + \frac{1}{3}\right)\left(\frac{\partial u_1''}{\partial x_1} + \frac{\partial u_2''}{\partial x_2} + \delta u_3''\right) = 0, \tag{5.18}$$

where the operator Δ_2 has the form

$$\Delta_2 = \frac{\partial^2}{\partial x_1^2} + \frac{\partial^2}{\partial x_2^2} - \delta^2.$$

System (5.18) defines the differential eigenvalue problem with the spectral parameter Re.

The velocity pulsation vector $\boldsymbol{u}''$ can be represented as

$$\boldsymbol{u}'' \equiv \left(u_1'',\, u_2'',\, u_3''\right) = \boldsymbol{v}\exp\left(i\alpha x_1\right), \tag{5.19}$$

where $\boldsymbol{v} = (u(x_2),\ v(x_2),\ w(x_2))$ is the vector of the perturbation amplitudes, α is the absolute value of the projection of the wave vector onto the x_1 coordinate axis, and i is the imaginary unit. Substitution of Eq. (5.19) into the Euler-Lagrange equations (5.18) leads to the following system of differential equations for the amplitudes u, v, and w:

$$\frac{d^2u}{dx_2^2} + i\alpha\left(\alpha_1 + \frac{1}{3}\right)\frac{dv}{dx_2} - \left[\alpha^2\left(\alpha_1 + \frac{4}{3}\right) + \delta^2\right]u - \frac{\mathrm{Re}}{2}v + i\alpha\delta\left(\alpha_1 + \frac{1}{3}\right)w = 0,$$

$$\left(\alpha_1 + \frac{4}{3}\right)\frac{d^2v}{dx_2^2} + i\alpha\left(\alpha_1 + \frac{1}{3}\right)\frac{du}{dx_2} + \delta\left(\alpha_1 + \frac{1}{3}\right)\frac{dw}{dx_2} - \frac{\mathrm{Re}}{2}u - \left(\alpha^2 + \delta^2\right)v = 0,$$

$$\frac{d^2w}{dx_2^2} - \delta\left(\alpha_1 + \frac{1}{3}\right)\frac{dv}{dx_2} - i\alpha\delta\left(\alpha_1 + \frac{1}{3}\right)u - \left[\delta^2\left(\alpha_1 + \frac{4}{3}\right) + \alpha^2\right]w = 0,$$

$$u\Big|_{x_2=\pm 1/2} = v\Big|_{x_2=\pm 1/2} = w\Big|_{x_2=\pm 1/2} = 0. \tag{5.20}$$

The spectral problem (5.20) has the following properties.

(*i*) All eigenvalues Re of problem (5.20) are real, which follows from the energy identity for this system. The latter is obtained by multiplying Eqs. (5.20) by complex-conjugate functions u^*, v^*, and w^*, summing them up, and integrating over the interval $x_2 \in [-1;\ 1]$. In view of homogeneous boundary conditions one has the expression

$$\mathrm{Re}\int_{-1}^{1}(v_r u_r + v_i u_i)\,dx_2 = \int_{-1}^{1}\left[\left|\frac{du}{dx_2}\right|^2 + \left(\alpha_1 + \frac{4}{3}\right)\left|\frac{dv}{dx_2}\right|^2 + \left|\frac{dw}{dx_2}\right|^2\right]dx_2$$

$$+\int_{-1}^{1}\left[\left(\alpha^2\left(\alpha_1 + \frac{4}{3}\right) + \delta^2\right)|u|^2 + \left(\alpha^2 + \delta^2\right)|v|^2\right.$$

$$\left.+\left(\delta^2\left(\alpha_1 + \frac{4}{3}\right) + \alpha^2\right)|w|^2\right]dx_2$$

$$+2\left(\alpha_1 + \frac{1}{3}\right)\int_{-1}^{1}\left[\delta\alpha\left(u_r w_i - u_i w_r\right) + \alpha\left(v_r\frac{du_i}{dx_2} + u_r\frac{dv_i}{dx_2}\right)\right.$$

$$\left.+\delta\left(w_r\frac{dv_r}{dx_2} + w_i\frac{dv_i}{dx_2}\right)\right]dx_2.$$

Here, the subscripts r and i indicate the real and imaginary parts of the corresponding complex-valued functions. The reality of the spectral parameter Re is determined by the reality of all terms of the resultant identity. At the same time the quadratic form determined by the energy identity is not positively determined. This means that the eigenvalues can also be negative; therefore, the minimum absolute eigenvalue min |Re| should be sought in the calculations.

(*ii*) The spectrum of the eigenvalues, $\mathrm{Re}(\alpha,\ \delta)$, is symmetric with respect to the axes $\alpha = 0$ and $\delta = 0$ on the plane of the wavenumbers $(\alpha,\ \delta)$. Indeed, it follows from the equations of system (5.20) that each eigenvalue, $\mathrm{Re}(\alpha,\ \delta)$, with the eigenfunctions u, v, and w corresponds to an equal eigenvalue, $\mathrm{Re}(-\alpha,\ \delta)$, with the eigenfunctions u^*, v^*, and w^*. The same eigenvalue, $\mathrm{Re}(\alpha,\ \delta)$, also corresponds to a pair of wave numbers, $(\alpha,\ -\delta)$, and to a set of eigenfunctions u, v, and $-w$.

One can note that system (5.20) is not reduced to a lower-order system by a linear change of variables, as in linear stability theory (cf. [12]); therefore, analytical results can be obtained only in particular cases, which are considered below.

Constant Mode $\alpha = \delta = 0$. In this case, system (5.20) becomes

$$\frac{d^2u}{dx_2^2} - \frac{\mathrm{Re}}{2}v = 0, \quad \left(\alpha_1 + \frac{4}{3}\right)\frac{d^2v}{dx_2^2} - \frac{\mathrm{Re}}{2}u = 0, \quad \frac{d^2w}{dx_2^2} = 0,$$

$$u\big|_{x_2=\pm 1/2} = v\big|_{x_2=\pm 1/2} = w\big|_{x_2=\pm 1/2} = 0. \tag{5.21}$$

The third equation of system (5.21) is integrated separately and has the general solution

$$w = c_1x_2 + c_2,$$

which vanishes identically under zero boundary conditions.

The characteristic equation of the abridged system (5.21) becomes

$$\lambda^4 - (\mathrm{Re}/2)^2(\alpha_1 + 4/3)^{-1} = 0.$$

The roots of this equation are

$$\lambda_{1,2} = \pm a, \qquad \lambda_{3,4} = \pm ia, \qquad a = \sqrt{\mathrm{Re}/2}(\alpha_1 + 4/3)^{-1/4}.$$

The general solution of the abridged system (5.21) is written as

$$\boldsymbol{V} = c_1\boldsymbol{V}_1\exp(ax_2) + c_2\boldsymbol{V}_2\exp(-ax_2) + c_3\boldsymbol{V}_3\cos(ax_2) + c_4\boldsymbol{V}_4\sin(ax_2),$$

where $\boldsymbol{V} = (u, v)$; $\boldsymbol{V}_k = (u_k, v_k)$ $(k = 1, 2, 3, 4)$ are eigenvectors. Using the homogeneous boundary conditions, we obtain $\boldsymbol{V}_1 = \boldsymbol{V}_2 \equiv 0$; nontrivial solutions are possible in two cases:

$$\boldsymbol{V}_3 \neq 0, \qquad \boldsymbol{V}_4 = 0, \qquad \cos(a/2) = 0 \tag{5.22}$$

or

$$\boldsymbol{V}_3 = 0, \qquad \boldsymbol{V}_4 \neq 0, \qquad \sin(a/2) = 0. \tag{5.23}$$

As a result from conditions (5.22) and (5.23), it follows that the eigenvalue spectra have the following form, respectively:

$$\mathrm{Re}^{(0)}_{cr,n} = 2\pi^2(2n-1)^2(\alpha_1 + 4/3)^{1/2},$$

$$\mathrm{Re}^{(0)}_{s,n} = 8\pi^2n^2(\alpha_1 + 4/3)^{1/2}, \quad n = 1, 2, 3, \ldots .$$

The critical value of the Reynolds number $\mathrm{Re}^{(0)}_{cr}$ is determined as the minimum value of the sets $\mathrm{Re}^{(0)}_{1,n}$ and $\mathrm{Re}^{(0)}_{2,n}$:

$$\mathrm{Re}^{(0)}_{cr} = \min_{n\in \mathrm{N}}\left(\mathrm{Re}^{(0)}_{cr,n}, \mathrm{Re}^{(0)}_{s,n}\right) = 2\pi^2(\alpha_1 + 4/3)^{1/2}.$$

Longitudinal Long-wave Modes $\alpha \ll 1$ *and* $\delta = 0$. For $\delta = 0$ system (5.20) reduces to the system

$$\frac{d^2u}{dx_2^2} + i\alpha\left(\alpha_1 + \frac{1}{3}\right)\frac{dv}{dx_2} - \alpha^2\left(\alpha_1 + \frac{4}{3}\right)u - \frac{\mathrm{Re}}{2}v = 0,$$

$$\left(\alpha_1 + \frac{4}{3}\right)\frac{d^2v}{dx_2^2} + i\alpha\left(\alpha_1 + \frac{1}{3}\right)\frac{du}{dx_2} - \frac{\mathrm{Re}}{2}u - \alpha^2 v = 0,$$

$$\frac{d^2w}{dx_2^2} - \alpha^2 w = 0,$$

$$u\big|_{x_2=\pm 1/2} = v\big|_{x_2=\pm 1/2} = w\big|_{x_2=\pm 1/2} = 0. \tag{5.24}$$

The equation for the transverse component in (5.24) is integrated separately and has the general solution

$$w = c_1 \exp(\alpha x_2) + c_2 \exp(-\alpha x_2)\,.$$

Substitution of this solution into the zero boundary conditions for w yields the following homogeneous system for arbitrary constants:

$$c_1 \exp(\alpha/2) + c_2 \exp(-\alpha/2) = 0,$$

$$c_1 \exp(-\alpha/2) + c_2 \exp(\alpha/2) = 0.$$

It follows that, for $\alpha \neq 0$, the solution $w \equiv 0$.

For the thus abridged system (5.24) the characteristic equation becomes an incomplete algebraic equation of fourth power, which can be written in standard form [13]

$$\lambda^4 + p\lambda^2 + q\lambda + r = 0, \tag{5.25}$$

where

$$p = -2\alpha^2, \quad q = i\alpha\mathrm{Re}(1 + 3\alpha_1)/(4 + 3\alpha_1),$$

$$r = \beta^4 - 3\mathrm{Re}^2/[4(4 + 3\alpha_1)].$$

The roots of Eq. (5.25) are calculated through the roots of the resolvent cubic equation, written in reduced form as follows

$$z^3 + p_1 z + q_1 = 0,$$

$$p_1 = \frac{3\mathrm{Re}^2}{4 + 3\alpha_1} - \frac{16}{3}\alpha^4,$$

$$q_1 = \frac{\mathrm{Re}}{4+3\alpha_1} 2(9\alpha_1^2 + 18\alpha_1 + 17)\alpha^2 - \frac{128}{27}\alpha^6.$$

The discriminant of the cubic resolvent is

$$D = (p_1/3)^3 + (q_1/2)^2 > 0.$$

From this it follows that, for arbitrary α, Eq. (5.25) has two real and two complex conjugate roots.

Consider the long-wave approximation assuming that $\alpha \ll 1$. The roots of Eq. (5.25) are calculated from the formulas [13]

$$\lambda_1 = \frac{1}{2}\left(\sqrt{z_1 + 4\alpha^2/3} + \sqrt{z_2 + 4\alpha^2/3} + \sqrt{z_3 + 4\alpha^2/3}\right),$$

$$\lambda_2 = \frac{1}{2}\left(\sqrt{z_1 + 4\alpha^2/3} - \sqrt{z_2 + 4\alpha^2/3} - \sqrt{z_3 + 4\alpha^2/3}\right),$$

$$\lambda_3 = \frac{1}{2}\left(-\sqrt{z_1 + 4\alpha^2/3} + \sqrt{z_2 + 4\alpha^2/3} - \sqrt{z_3 + 4\alpha^2/3}\right),$$

$$\lambda_4 = \frac{1}{2}\left(-\sqrt{z_1 + 4\alpha^2/3} - \sqrt{z_2 + 4\alpha^2/3} + \sqrt{z_3 + 4\alpha^2/3}\right),$$

where z_k $(k = 1, 2, 3)$ are roots of the resolvent cubic equation. The roots of the characteristic equation (5.25) are expressed, to within terms of order $O(\alpha^4)$, as follows:

$$\lambda_{1,2} = \pm a_1 + ib, \quad \lambda_{3,4} = \pm i(a_2 \mp b),$$

$$a_{1,2} = \sqrt{\frac{\mathrm{Re}}{2}}\left(\alpha_1 + \frac{4}{3}\right)^{-1/4}\left[1 \pm \frac{\alpha^2(7+3\alpha_1)^2}{36\mathrm{Re}}\left(\alpha_1 + \frac{4}{3}\right)^{-1/2}\right],$$

$$b = \frac{\alpha(1+3\alpha_1)}{6}\left(\alpha_1 + \frac{4}{3}\right)^{-1/2}.$$

In the abridged system (5.24), since the coefficients are complex-valued, the amplitudes of the velocity pulsations u and v are expressed in terms of the real part of the general solution of this system:

$$Real(\boldsymbol{V}) = Real\left(\sum_{k=1}^{4} c_k \boldsymbol{V}_k \exp(\lambda_k)\right), \tag{5.26}$$

where $\boldsymbol{V}_k = (u_{k1} + iu_{k2}, v_{k1} + iv_{k2})$ are complex-valued eigenvectors.

Considering each term in (5.26) separately and taking into account the homogeneous boundary conditions for the component of the eigenvectors for $k = 1, 2$, one obtains the systems

$$q_{k1}\cos(b/2) - q_{k2}\sin(b/2) = 0, \quad q_{k1}\cos(b/2) + q_{k2}\sin(b/2) = 0,$$

$$q_{kj} = (u_{kj}, v_{kj}).$$

These systems have nontrivial solutions if $\sin b = 0$, which is ruled out in the case $\alpha \neq 0$. From this, we have $\boldsymbol{V}_1 = \boldsymbol{V}_2 = 0$. Similarly, for the eigenvector components for $k = 3,\ 4$, one has homogeneous systems of the form

$$q_{k1}\cos\left(\frac{a_2 \mp b}{2}\right) \mp q_{k2}\sin\left(\frac{a_2 \mp b}{2}\right) = 0,$$

$$q_{k1}\cos\left(\frac{a_2 \mp b}{2}\right) \pm q_{k2}\sin\left(\frac{a_2 \mp b}{2}\right) = 0,$$

$$q_{kj} = (u_{kj}, v_{kj}),$$

where the upper signs correspond to the case $k = 3$, and the lower signs to $k = 4$. These systems have nontrivial solutions if the following conditions are satisfied:

$$\sin(a_2 \mp b) = 0. \tag{5.27}$$

Using (5.27), one obtains the following equations for the eigenvalues Re:

$$x^2 - p_{\pm}x - s = 0. \tag{5.28}$$

Here

$$x = \sqrt{\mathrm{Re}/2}(\alpha_1 + 4/3)^{-1/4}, \quad s = \alpha^2(7 + 3\alpha_1)^2/[24(4 + 3\alpha_1)],$$

$$p_{\pm} = \pi n[1 \pm \alpha(1 + 3\alpha_1)(\alpha_1 + 4/3)^{-1/2}/(6\pi n)],$$

$$n = 1,\ 2,\ 3,\ \ldots, \tag{5.29}$$

where the upper sign corresponds to the first condition in (5.27).

The roots of the quadratic equation (5.28), to within terms of order $O(\alpha^2)$, are given by

$$x_1 = \pi n\left[1 \pm \frac{\alpha(1 + 3\alpha_1)}{6\pi n}\left(\alpha_1 + \frac{4}{3}\right)^{-1/2}\right]$$

$$+\frac{\alpha^2\pi n}{4}\left[\frac{(1 + 3\alpha_1)^2}{(6\pi n)^2}\left(\alpha_1 + \frac{4}{3}\right)^{-1} + \frac{(7 + 3\alpha_1)^2}{18(\pi n)^2}\left(\alpha_1 + \frac{4}{3}\right)^{-1}\right],$$

$$x_2 = -\frac{\alpha^2 \pi n}{4}\left[\frac{(1+3\alpha_1)^2}{(6\pi n)^2}\left(\alpha_1+\frac{4}{3}\right)^{-1}+\frac{(7+3\alpha_1)^2}{18(\pi n)^2}\left(\alpha_1+\frac{4}{3}\right)^{-1}\right].$$

The root x_2 is not considered further since it corresponds to the out-of-order dependence $\mathrm{Re} \sim O(\alpha^4)$, which was neglected above in the expressions for the roots λ_k of the characteristic equation.

The spectra of the eigenvalues Re are given, to within terms of order $O(\alpha^3)$, for the root x_1, by the relations

$$\mathrm{Re}_n^{(\alpha)} = 2\pi^2 n^2\left(\alpha_1+\frac{4}{3}\right)^{1/2}\left\{\left[1 \pm \frac{\alpha(1+3\alpha_1)}{6\pi n}\left(\alpha_1+\frac{4}{3}\right)^{-1/2}\right]^2\right.$$
$$\left.+\frac{\alpha^2}{2}\left[\frac{(1+3\alpha_1)^2}{(6\pi n)^2}\left(\alpha_1+\frac{4}{3}\right)^{-1}+\frac{(7+3\alpha_1)^2}{18(\pi n)^2}\left(\alpha_1+\frac{4}{3}\right)^{-1}\right]\right\},$$
$$n = 1,\ 2,\ 3, \ldots, \tag{5.30}$$

where the plus sign corresponds to the first condition in (5.27).

From (5.30) it follows that the minimum Reynolds number $\mathrm{Re}_{cr}^{(\alpha)}$ for long-wave longitudinal modes is

$$\mathrm{Re}_{cr}^{(\alpha)} = 2\pi^2\left(\alpha_1+\frac{4}{3}\right)^{1/2}\left[1-\frac{\alpha(1+3\alpha_1)}{3\pi}\left(\alpha_1+\frac{4}{3}\right)^{-1/2}\right.$$
$$\left.+\frac{\alpha^2}{72\pi^2}(45\alpha_1^2+102\alpha_1+101)\left(\alpha_1+\frac{4}{3}\right)^{-1}\right]. \tag{5.31}$$

Transverse Long-wave Modes $\alpha = 0$ *and* $\delta \ll 1$. Investigation of these modes is of interest since, for an incompressible Couette flow, the critical Reynolds number closest to experimental values was obtained for a transverse mode [5]. For $\alpha = 0$ system (5.20) becomes

$$\frac{d^2u}{dx_2^2} - \delta^2 u - \frac{\mathrm{Re}}{2}v = 0,$$
$$\left(\alpha_1+\frac{4}{3}\right)\frac{d^2v}{dx_2^2} + \delta\left(\alpha_1+\frac{1}{3}\right)\frac{dw}{dx_2} - \frac{\mathrm{Re}}{2}u - \delta^2 v = 0,$$
$$\frac{d^2w}{dx_2^2} - \delta\left(\alpha_1+\frac{1}{3}\right)\frac{dv}{dx_2} - \delta^2\left(\alpha_1+\frac{4}{3}\right)w = 0,$$
$$u\big|_{x_2=\pm 1/2} = v\big|_{x_2=\pm 1/2} = w\big|_{x_2=\pm 1/2} = 0. \tag{5.32}$$

The characteristic equation of system (5.32) is written as

$$\lambda^6 - 3\delta^2\lambda^4 + \frac{3}{4}\left(4\delta^4 - \frac{\mathrm{Re}^2}{4+3\alpha_1}\right)\lambda^2 - \delta^2\left(\delta^4 - \frac{\mathrm{Re}^2}{4}\right) = 0. \tag{5.33}$$

The change $z = \lambda^2 - \delta^2$ transforms this equation to the reduced cubic equation

$$z^3 + p_1 z + q_1 = 0,$$

where

$$p_1 = -(\mathrm{Re}/2)^2(\alpha_1 + 4/3)^{-1},$$

$$q_1 = \delta^2\mathrm{Re}^2(1+3\alpha_1)(\alpha_1+4/3)^{-1}/12.$$

The discriminant of this equation is negative

$$D = (p_1/3)^3 + (q_1/2)^2 < 0$$

and for the character dependence $\mathrm{Re}(\alpha_1)$ remains negative even in the case $\delta \sim O(1)$. The reduced cubic equation has three real roots defined by the Cardano formulas [13, 14]

$$z_k = 2\xi^{1/3}\cos\left[(\varphi + 2k\pi)/3\right], \quad k = 0, 1, 2.$$

Here

$$\xi = \sqrt{-\left(\frac{p_1}{3}\right)^3} = \left[\frac{\mathrm{Re}}{2\sqrt{3}}\left(\alpha_1 + \frac{4}{3}\right)^{-1/2}\right]^3,$$

$$\cos\varphi = -\frac{q_1}{2\xi} = -\frac{\delta^2\sqrt{3}}{\mathrm{Re}}(1+3\alpha_1)\left(\alpha_1 + \frac{4}{3}\right)^{1/2}.$$

Retaining terms of order not higher than $O(\delta^2)$ in the expressions for z_k, we obtain the following expressions for the roots of the characteristic equation (5.33):

$$\lambda_{1,2} = \pm\sqrt{z_0 + \delta^2} = \pm\delta(\alpha_1 + 4/3)^{1/2},$$

$$\lambda_{3,4} = \pm\sqrt{z_1 + \delta^2} = \pm\sqrt{\mathrm{Re}(\alpha_1 + 4/3)^{-1/2}/2 + \delta^2(5 - 3\alpha_1)/6},$$

$$\lambda_{5,6} = \pm\sqrt{z_2 + \delta^2} = \pm i\sqrt{\mathrm{Re}(\alpha_1 + 4/3)^{-1/2}/2 - \delta^2(5 - 3\alpha_1)/6}.$$

Thus, the first four roots λ_k ($k = 1, 2, 3, 4$) are real, and the roots $\lambda_{5,6}$ are complex conjugate, purely imaginary.

Because the velocity pulsation amplitude vector $\boldsymbol{v}$ is real, it is expressed in terms of the real part of the general solution of system (5.32):

$$\boldsymbol{v} = Real(\boldsymbol{V}) = Real\left(\sum_{k=1}^{4} c_k \boldsymbol{V}_k \exp(\lambda_k)\right), \tag{5.34}$$

where $\boldsymbol{V}_k = (u_{k1} + iu_{k2}, v_{k1} + iv_{k2}, w_{k1} + iw_{k2})$ are complex-valued eigenvectors.

For each term in (5.34), in view of the homogeneous boundary conditions for the amplitudes, it follows that the eigenvectors corresponding to the real roots are zero:

$$\boldsymbol{V}_1 = \boldsymbol{V}_2 = \boldsymbol{V}_3 = \boldsymbol{V}_4 = 0.$$

For the eigenvector components for $k = 5, 6$, the following homogeneous systems hold:

$$q_{k1}\cos(|\lambda_k|/2) - q_{k2}\sin(|\lambda_k|/2) = 0,$$

$$q_{k1}\cos(|\lambda_k|/2) + q_{k2}\sin(|\lambda_k|/2) = 0,$$

$$q_{kj} = (u_{kj}, v_{kj}, w_{kj}).$$

Nontrivial solutions of these systems exist for $\sin|\lambda_k| = 0$. In this case, the eigenvalue spectrum has the form

$$\mathrm{Re}_n^{(\delta)} = 2(\alpha_1 + 4/3)^{1/2}[\pi^2 n^2 + \delta^2(5/3 - \alpha_1)/2], \quad n = 1, 2, 3, \ldots .$$

Consequently, the critical Reynolds number, as the minimal eigenvalue of the set of eigenvalues for transverse modes, is

$$\mathrm{Re}_{cr}^{(\delta)} = 2\pi^2(\alpha_1 + 4/3)^{1/2}[1 + \delta^2(5/3 - \alpha_1)/(2\pi^2)].$$

Asymptotic estimates of different modes derived in the main order contain the characteristic dependence

$$\mathrm{Re}_{cr} \sim \sqrt{\alpha_1 + 4/3}.$$

This implies that, at ratios of the bulk to shear viscosity (parameter α_1) realistic for diatomic gases, the critical Reynolds number can considerably increase with increasing bulk viscosity. The estimates are consistent with data on the effect of bulk viscosity on the stability of boundary layers on a plate obtained within the framework of linear theory [3, 15].

The asymptotics considered are long-wave approximations. This suggests that the obtained relation describes the effect of bulk viscosity on large-scale vortex structures.

5.1.3 *Results of Numerical Calculation of the Spectral Problem*

The spectral problem (5.20) was solved numerically in the Matlab environment with arbitrary values of the wavenumbers α and δ. The method of collocations described in Sect. 3.2 was used. Here the Gauss–Lobatto points $x_{2,k} = \cos(\pi k/N)$, $k = 0, 1, \ldots, N$, in which the N-power Chebyshev polynomial $T_N(x_2)$ possesses extremums on the segment $[-1;\ 1]$, were chosen as collocation (interpolation) nodes. The first-order differential operators involved in the spectral problem are approximated on this template by a matrix of collocation derivatives D_N^1 [16] of size $(N+1) \times (N+1)$ (see (3.31)). The elements of the lth row of the matrix D_N^1 are coefficients of the difference approximation of the first derivative in the lth node of collocation on the template $\{x_{2,k}\}$. The second-order differential operators are approximated by the superposition [16, 17]

$$D_N^2 = D_N^1 D_N^1. \tag{5.35}$$

In terms of these approximations, problem (5.20) reduces to a generalized eigenvalue problem (linear matrix bundle) with respect to the spectral parameter $\lambda = \mathrm{Re}/2$:

$$\sum_{j=0}^{3N+2} \left[G_{kj} - \lambda F_{kj}\right] q_j = 0, \quad k = 0, 1, \ldots, 3N+2. \tag{5.36}$$

Here, the vector of the unknowns $\boldsymbol{q}$ of size $3(N+1)$ consists of the values of the eigenfunctions in the collocation nodes

$$\boldsymbol{q} = (u_0, u_1, \ldots, u_N,\ v_0, v_1, \ldots, v_N,\ w_0, w_1, \ldots, w_N),$$

the matrices G and F of size $3(N+1) \times 3(N+1)$ are calculated with a special Matlab procedure by the formulas

$$G = A \otimes D_N^2 + B \otimes D_N^1 + C \otimes I_N, \quad F = K \otimes I_N,$$

the symbol "$\otimes$" means direct (tensor) multiplication of the matrices [14], I_N is a unit matrix of size $(N+1) \times (N+1)$, and A, B, C, and K are matrices of size 3×3:

$$A = \begin{pmatrix} 1 & 0 & 0 \\ 0 & \alpha_1 + 4/3 & 0 \\ 0 & 0 & 1 \end{pmatrix},$$

$$B = \begin{pmatrix} 0 & i\alpha(\alpha_1 + 1/3) & 0 \\ i\alpha(\alpha_1 + 1/3) & 0 & \delta(\alpha_1 + 1/3) \\ 0 & -\delta(\alpha_1 + 1/3) & 0 \end{pmatrix},$$

$$C = \begin{pmatrix} -\left[\alpha^2(\alpha_1 + 4/3) + \delta^2\right] & 0 & i\alpha\delta(\alpha_1 + 1/3) \\ 0 & -(\alpha^2 + \delta^2) & 0 \\ -i\alpha\delta(\alpha_1 + 1/3) & 0 & -\left[\delta^2(\alpha_1 + 4/3) + \beta^2\right] \end{pmatrix},$$

$$K = \begin{pmatrix} 0\,1\,0 \\ 1\,0\,0 \\ 0\,0\,0 \end{pmatrix}.$$

The homogeneous boundary conditions for system (5.36) are taken into account implicitly, via the operator D_N^1 [16]. Exactly, they are implemented through replacing the matrices D_N^k $(k = 1, 2)$ by bordered matrices of size $(N - 1) \times (N - 1)$. The latter are obtained if the following conditions are satisfied:

$$D_{0,j}^k = D_{N+1,j}^k = 0, \quad D_{l,0}^k = D_{l,N+1}^k,$$

$$l,\ j = 0,\ \ldots,\ N + 1, \quad k = 1,\ 2. \tag{5.37}$$

To find all eigenvalues and functions of the generalized spectral problem (5.36), we used as in Sect. 3.2 the Matlab procedure implementing the QZ-algorithm [18].

By using this procedure for fixed values of the parameter α_1 (bulk viscosity) and each pair of wavenumbers $(\alpha,\ \delta)$, we obtained a set of $N + 1$ eigenvalues, which include the Reynolds number with the minimum absolute value $\mathrm{Re}(\alpha,\ \delta) = 2\,|\lambda_{\min}(\alpha,\ \delta)|$, and calculated the corresponding eigenfunctions u, v, and w. The value of the critical Reynolds number Re_{cr} for this α_1 was taken to be the minimum value of Re in the entire range of wavenumbers $\mathrm{Re}(\alpha,\ \delta)$:

$$\mathrm{Re}_{cr} = \min_{(\alpha,\ \delta)} \left[\mathrm{Re}(\alpha,\ \delta)\right].$$

The spectra of the eigenvalues $\lambda(\alpha_1, \alpha, \delta)$ were calculated for the values of the parameter $\alpha_1 = 0$, 0.5, 1, 1.5, and 2 in the wavenumber range $\alpha = 0$–5, $\delta = 0$–5. The calculation accuracy was checked by calculations in the range $\alpha = -5$–5, $\delta = -5$–5 using spectrum symmetry. The steps of wavenumber variations were chosen to be $h_\alpha = h_\delta = 0.1$. In most calculations the number of collocation nodes was taken to be $N + 1 = 50$. To check the accuracy of calculating the eigenvalues, the number of collocation nodes $N + 1 = 32$–100 was also varied.

Examples of the calculated dependencies $\mathrm{Re}(\alpha,\ \delta)$ for different values of the bulk viscosity α_1 are shown in Fig. 5.1. The calculations show that, for all fixed values of α_1, the minimum absolute values of $\mathrm{Re}(\alpha,\ \delta) = 2\,|\lambda_{\min}(\alpha,\ \delta)|$ are reached on the axis $\alpha \neq 0$ (at $\delta = 0$) in the plane of the wavenumbers $(\alpha,\ \delta)$. The spectral parameter is plotted in Fig. 5.2 as a function of the wavenumber $\mathrm{Re}(\alpha)$ for $\delta = 0$ and for chosen values of the bulk viscosity α_1. By such a way we obtain neutral stability curves, in fact. Because along these curves $dE_t(t)/dt = 0$. The dot-and-dashed curve connects the absolute minimums on the curves $\mathrm{Re}_{cr}(\alpha_1)$ parameterized with respect to α_1. The values of these quantities versus the parameter α_1 are given in Table 5.1.

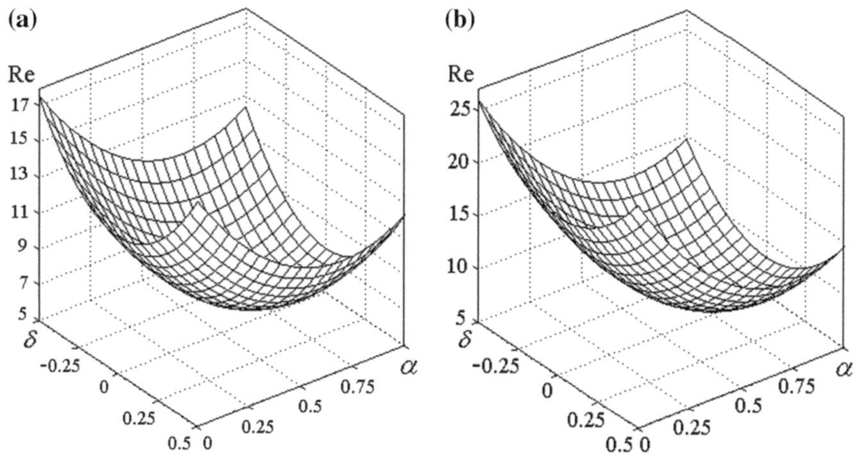

Fig. 5.1 Spectral parameter versus the wavenumber Re(α, δ). **a** is $\alpha_1 = 0.5$. **b** is $\alpha_1 = 1.5$

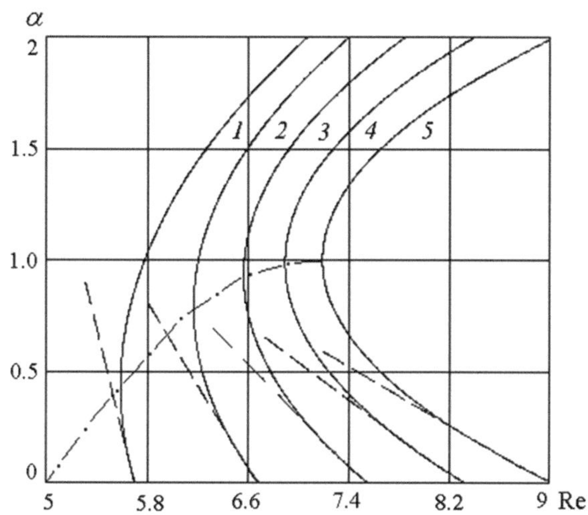

Fig. 5.2 Dependencies Re(α) for $\delta = 0$ (neutral stability curves) at $\alpha_1 = 0$ (*1*), 0.5 (*2*), 1 (*3*), 1.5 (*4*), and 2 (*5*). The *dashed curves* are the asymptotic functions $\mathrm{Re}_{cr}^{(\alpha)}$ (5.31). The *dot-and-dashed curve* is the critical Reynolds number versus the wavenumber $\mathrm{Re}_{cr}(\alpha)$

Analysis of the calculated curves shows that the critical Reynolds numbers Re_{cr} and the corresponding values of the wavenumber α increase with increasing parameter α_1. Thus, as the parameter α_1 increases in the interval typical for diatomic gases, the increase in the critical Reynolds number may reach 30%.

In the case of long-wave disturbances, the asymptotic values of $\mathrm{Re}_{cr}^{(\alpha)}$ determined by Eq. (5.31) are substantially greater than the corresponding calculated critical values of the Reynolds number Re(α, α_1) and become close to the latter only at $\alpha < 0.25$. As one can see formula (5.31) presents an asymptotics of lower reach of neutral stability curves.

The dependencies of the real (u_r and v_r) and imaginary (u_i and v_i) parts of the eigenfunctions u and v corresponding to the critical Reynolds numbers $\mathrm{Re}_{cr}(\alpha_1)$ on

Table 5.1 Critical Reynolds number $\mathrm{Re}_{cr}(\alpha_1)$ as a function of the wavenumber α for $\delta = 0$

α_1	Re_{cr}	α
0.0	5.585	0.431
0.5	6.170	0.782
1.0	6.565	0.929
1.5	6.892	0.982
2.0	7.189	0.993

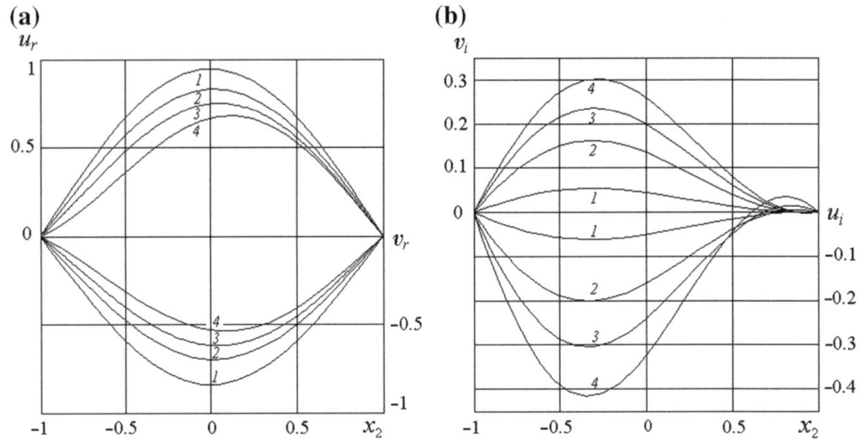

Fig. 5.3 Eigenfunctions u_r, v_r, u_i, and v_i corresponding to the critical Reynolds numbers $\mathrm{Re}_{cr}(\alpha)$ versus the x_2 coordinate for $\alpha = 0$ (*1*), 0.5 (*2*), 1 (*3*), and 2 (*4*). **a** shows eigenfunctions u_r and v_r. **b** shows eigenfunctions u_i and v_i

the coordinate x_2 are shown in Fig. 5.3. It was demonstrated in [5] that the eigenfunctions w of system (5.20) at $\delta = 0$ are equal to zero.

As the disturbances are periodic with respect to the streamwise coordinate x_1, the most dangerous disturbances determined on the basis of the energy theory can be interpreted as a chain of vortices whose axes are normal to the flow plane.

It should be noted that the critical Reynolds numbers are reached in this case on the streamwise modes $\alpha \neq 0$ and $\delta = 0$. At the same time, in all calculations of stability of plane incompressible flows within the framework of the energy theory available to the authors (see [5, 12] and the references therein), the minimum Reynolds numbers were obtained for the transverse modes $\alpha = 0$ and $\delta \neq 0$. The values of these minimum Reynolds numbers are substantially higher than those obtained in the present work. Therefore, the most dangerous disturbances were two-dimensional vortices with the axes parallel to the flow. The question of inconsistency of these results with the known Squire's theorem [9, 12] remains open, though the significance of the latter should not be overestimated.

In particular a plane Couette flow was considered in [5] on the basis of the Oberbek–Boussinesq equations for an inhomogeneous incompressible fluid. The same method was used in [4] to study the stability of the Poiseuille flow in pipes and channels on the basis of the Navier–Stokes model. The problem on a conditional extremum of the energy functional on the subspace of solenoidal fields was considered in both cases.

It is known that compressibility under subsonic Mach numbers cannot exert any significant effect on the Couette flow properties from the physical viewpoint. At the same time the variational problem (5.20) differs substantially from the corresponding problem for incompressible flow [5, 12]. Therefore, the above-mentioned difference can be caused either by the additional condition of a solenoidal flow or by the assumption about character of disturbance dependence on the transverse coordinate (5.16). To check the latter assumption with use of the same Matlab procedure calculations [12] of Re_{cr} the incompressible Couette flow were repeated with separation of the transverse coordinate defined by (5.16). The obtained results coincided with ones given in [12].

5.2 Energy Stability Analysis of a Plane Vibrationally Excited Flow. Effect of a Vibrational Relaxation

The influence of intense thermal excitation on the critical Reynolds number Re_{cr} is studied analytically and numerically on the basis of the energy stability theory in this section. The Couette flow of a vibrationally excited diatomic gas described by the system of equations of two-temperature gas dynamics is considered. In accordance with the physical concepts [2, 7, 12] (see also Chap. 1), this system is a commonly accepted mathematical model of vibrationally excited molecular gas flows when one can neglect by the molecular dissociation, excitation of the upper vibrational levels and anharmonism of vibrations.

5.2.1 Energy Balance Equation of Total Disturbances

The problem of stability of the Couette flow of a vibrationally excited diatomic gas is considered in a rectangular computational domain Ω with the Cartesian system (x_1, x_2, x_3) defined in previous section.

The characteristic scales for normalization are the same used in Sect. 5.1. In the dimensionless variables the system of equations of two-temperature aerodynamics is written as

$$\frac{\partial \rho}{\partial t} + \frac{\partial \rho u_i}{\partial x_i} = 0,$$

$$\rho\left(\frac{\partial u_i}{\partial t} + u_j\frac{\partial u_i}{\partial x_j}\right) = -\frac{\partial p}{\partial x_i} + \frac{1}{\mathrm{Re}}\frac{\partial^2 u_i}{\partial x_j^2} + \frac{1}{\mathrm{Re}}\left(\alpha_1 + \frac{1}{3}\right)\frac{\partial^2 u_j}{\partial x_i\,\partial x_j},$$

$$\rho\left(\frac{\partial T}{\partial t}+u_j\frac{\partial T}{\partial x_j}\right)+(\gamma-1)\rho T\frac{\partial u_i}{\partial x_i}=\frac{\gamma}{\mathrm{Re\,Pr}}\frac{\partial^2 T}{\partial x_i^2}+\frac{\rho\gamma_v(T_v-T)}{\tau}$$

$$+\frac{\gamma(\gamma-1)\mathrm{M}^2}{2\mathrm{Re}}\left[\left(\frac{\partial u_i}{\partial x_j}+\frac{\partial u_j}{\partial x_i}\right)^2+2\left(\alpha_1-\frac{2}{3}\right)\left(\frac{\partial u_i}{\partial x_i}\right)^2\right],$$

$$\gamma_v\rho\left(\frac{\partial T_v}{\partial t}+u_j\frac{\partial T_v}{\partial x_j}\right)=\frac{\gamma\alpha_2}{\mathrm{Re\,Pr}}\frac{\partial^2 T_v}{\partial x_i^2}-\frac{\gamma_v\rho(T_v-T)}{\tau},$$

$$\gamma\mathrm{M}^2 p=\rho T,\quad i=1,2,3,\quad j=1,2,3, \tag{5.38}$$

where ρ, u_i, p, T, T_v, and τ are the density, velocity vector components, pressure, static (translational) and vibrational temperatures of the gas, and vibrational relaxation time, respectively. Hereinafter, summation over repeated indices is applied. The specific heats, dissipative coefficients, and time τ in system (5.38) are assumed to be constant and independent of the static and vibrational temperatures of the flow.

The parameters involved in Eqs. (5.38) are determined as follows: $\alpha_1=\eta_b/\eta$ (η_b and η are the bulk and shear viscosities), $\alpha_2=\lambda_v/(\lambda_t+\lambda_r)$ (λ_t, λ_r, and λ_v are the thermal conductivity coefficients that describe energy transfer by translational and rotational degrees of freedom of molecules, and also by vibrational energy quanta), $\gamma_v=c_{V,v}/(c_{V,t}+c_{V,r})$ is a parameter characterizing the degree of nonequilibrium of the vibrational mode ($c_{V,t}$, $c_{V,r}$, and $c_{V,v}$ are the specific heats at constant volume determining the energy intensity of translational, rotational, and vibrational modes, respectively), $\gamma=(c_V+R)/c_V$ is the ratio of specific heats ($c_V=c_{V,t}+c_{V,r}$, R is the gas constant), and $\mathrm{Re}=U_0L\rho_0/\eta$, $\mathrm{M}=U_0/\sqrt{\gamma RT_0}$, and $\mathrm{Pr}=\eta(c_V+R)/(\lambda_t+\lambda_r)$ are the Reynolds, Mach, and Prandtl numbers of the flow, respectively.

System (5.38) describes a typical situation in aerodynamics, where the characteristic times of microscopic processes of energy exchange between various degrees of freedom of gas molecules are estimated by the system of inequalities [2, 7]

$$\tau_{tt}\sim\tau_{rt}\ll\tau_{vv}\ll\tau_{vt}\sim\tau_0.$$

In this case the translational and rotational degrees of freedom of molecules with small commensurable relaxation times $\tau_{tt}\sim\tau_{rt}$ at times of the order of the characteristic time of the flow τ_0 form a quasi-equilibrium thermostat with the flow static temperature T. A quasi-equilibrium distribution with the vibrational temperature T_v is established in the subsystem of vibrational energy levels after the time interval τ_{vv}. The energy exchange between the vibrational mode and quasi-equilibrium degrees of freedom of gas molecules is described by the Landau-Teller relaxation equation with the characteristic time $\tau_{vt}\equiv\tau$.

This presentation allows us to reduce the number of independent parameters in system (5.38) as follows. Using the Eucken relations [2, 7]

$$\lambda_t=(5/2)\,\eta\,c_{V,t},\quad \lambda_r=(6/5)\,\eta\,c_{V,r},\quad \lambda_v=(6/5)\,\eta\,c_{V,v},$$

coupling the partial coefficient of thermal conductivity with the shear viscosity coefficient, we can write the expression for the parameter α_2 as

$$\alpha_2 = \frac{\lambda_v}{\lambda_t + \lambda_r} = \frac{(6/5)\, c_{V,\,v}}{(5/2)\, c_{V,\,t} + (6/5)\, c_{V,\,r}} = \frac{12\,\gamma_v\,(c_{V,\,t} + c_{V,\,r})}{25\, c_{V,\,t} + 12\, c_{V,\,r}}.$$

As the translational and rotational degrees of freedom of molecules are in the quasi-equilibrium state, their internal energy is uniformly distributed over the degrees of freedom. Therefore, the values of the corresponding specific heats are $c_{V,\,t} = (3/2)R$ and $c_{V,\,r} = R$. As a result, we obtain the expression $\alpha_2 = 20\gamma_v/33$.

As the stationary flow, we consider the Couette flow with a linear velocity profile and uniform distributions of density and temperatures:

$$\boldsymbol{U}_s(x_2) = (x_2, 0, 0), \quad T_s(x_2) = T_{v,s}(x_2) = \rho_s(x_2) = 1, \quad p_s(x_2) = 1/(\gamma \mathrm{M}^2).$$

The presentation of the instantaneous values of the hydrodynamic quantities of the disturbed flow in the form

$$\rho = 1 + \rho', \quad u_i = U_{s,i} + u_i', \quad T = 1 + T', \quad T_v = 1 + T_v', \quad p = 1/(\gamma \mathrm{M}^2) + p'$$

allows us to obtain equations for the main flow disturbances ρ', u_i', T', T_v', and p' without constraints on their amplitudes:

$$\frac{\partial \rho'}{\partial t} + u_i \frac{\partial \rho'}{\partial x_i} + \rho \frac{\partial u_i'}{\partial x_i} = 0\,, \tag{5.39}$$

$$\rho\left(\frac{\partial u_i'}{\partial t} + u_j' \frac{\partial u_i'}{\partial x_j} + U_{s,\,j} \frac{\partial u_i'}{\partial x_j} + u_j' \frac{\partial U_{s,\,i}}{\partial x_j}\right)$$

$$= -\frac{\partial p'}{\partial x_i} + \frac{1}{\mathrm{Re}} \frac{\partial^2 u_i'}{\partial x_j^2} + \frac{1}{\mathrm{Re}}\left(\alpha_1 + \frac{1}{3}\right) \frac{\partial^2 u_j'}{\partial x_i \partial x_j}\,, \tag{5.40}$$

$$\rho\left(\frac{\partial T'}{\partial t} + u_j' \frac{\partial T'}{\partial x_j} + U_{s,\,j} \frac{\partial T'}{\partial x_j}\right) + \gamma(\gamma - 1)\mathrm{M}^2 p \frac{\partial u_i'}{\partial x_i}$$

$$= \frac{\gamma}{\mathrm{Re\,Pr}} \frac{\partial^2 T'}{\partial x_i^2} + \frac{\gamma_v \rho (T_v' - T')}{\tau}$$

$$+ \frac{\gamma(\gamma - 1)\mathrm{M}^2}{2\mathrm{Re}} \left[\left(\frac{\partial u_i'}{\partial x_j} + \frac{\partial u_j'}{\partial x_i}\right)^2 + 2\left(\frac{\partial U_{s,\,i}}{\partial x_j} + \frac{\partial U_{s,\,j}}{\partial x_i}\right)\left(\frac{\partial u_i'}{\partial x_j} + \frac{\partial u_j'}{\partial x_i}\right)\right.$$

$$\left. + \left(\frac{\partial U_{s,\,i}}{\partial x_j} + \frac{\partial U_{s,\,j}}{\partial x_i}\right)^2 + 2\left(\alpha_1 - \frac{2}{3}\right)\left(\frac{\partial u_i'}{\partial x_i}\right)^2\right], \tag{5.41}$$

$$\gamma_v \rho \left(\frac{\partial T_v'}{\partial t} + u_j' \frac{\partial T_v'}{\partial x_j} + U_{s,j} \frac{\partial T_v'}{\partial x_j} \right) = \frac{20}{33} \frac{\gamma \gamma_v}{\mathrm{Re\,Pr}} \frac{\partial^2 T_v'}{\partial x_i^2} - \frac{\gamma_v \rho (T_v' - T')}{\tau}, \quad (5.42)$$

$$\gamma \mathrm{M}^2 p' = \rho T' + \rho', \quad i = 1, 2, 3, \quad j = 1, 2, 3. \quad (5.43)$$

It is assumed that the disturbances of the hydrodynamic variables satisfy periodic boundary conditions at $x_1 = \pm x_0/2$ and $x_3 = \pm z_0/2$ and take zero values on the impermeable boundaries $x_2 = \pm 1$. In our calculations the sizes of the domain Ω along the periodic (uniform) coordinates x_1 and x_3 were assumed to be equal to the disturbance wave length along the corresponding coordinate: $x_0 = 2\pi/\alpha$ and $z_0 = 2\pi/\delta$ (α and δ are the absolute values of the projections of the disturbance wave vector $\boldsymbol{k}$ onto the coordinate axes x_1 and x_3).

As the initial equation of the energy balance of disturbances for system (5.39)–(5.43), the equation of the energy balance of the kinetic energy of disturbances derived in (5.7) was considered. This equation has the form (5.9).

To take into account the specific features of the energy balance of disturbances in the vibrationally excited gas, one transforms the integral J_2 on the right side of Eq. (5.9). By virtue of the continuity equation (5.39), equation of state (5.43), equation of energy (5.41), and Landau–Teller equation (5.42), the expression for J_2 takes the form

$$J_2 = \int\limits_{\Omega} p' \frac{\partial u_i'}{\partial x_i} d\Omega = \frac{1}{\gamma \mathrm{M}^2} \int\limits_{\Omega} (\rho T' + \rho') \frac{\partial u_i'}{\partial x_i} d\Omega$$

$$= -\frac{1}{2} \frac{\partial}{\partial t} \int\limits_{\Omega} \left(\rho (T'^2 + \gamma_v T_v'^2) + \frac{\rho'^2}{\gamma \mathrm{M}^2} \right) d\Omega$$

$$- \int\limits_{\Omega} \left\{ \frac{\gamma}{\mathrm{Re\,Pr}} \left[\left(\frac{\partial T'}{\partial x_i} \right)^2 + \frac{20\gamma_v}{33} \left(\frac{\partial T_v'}{\partial x_i} \right)^2 \right] + \frac{\gamma_v \rho (T_v' - T')^2}{\tau} \right.$$

$$\left. + \left(\gamma - \frac{1 + \gamma \mathrm{M}^2}{\gamma \mathrm{M}^2} \right) \rho T' \frac{\partial u_i'}{\partial x_i} \right\} d\Omega + \frac{\gamma(\gamma - 1)\mathrm{M}^2}{2\mathrm{Re}} \int\limits_{\Omega} T' \left[\left(\frac{\partial u_i}{\partial x_j} + \frac{\partial u_j}{\partial x_i} \right)^2 \right.$$

$$\left. + 2\left(\alpha_1 - \frac{2}{3} \right) \left(\frac{\partial u_i'}{\partial x_i} \right)^2 \right] d\Omega - \int\limits_{\Omega} \left(\frac{\rho'}{\gamma \mathrm{M}^2} \frac{\partial \rho' u_i'}{\partial x_i} + (\gamma - 1) \rho T'^2 \frac{\partial u_i'}{\partial x_i} \right) d\Omega. \quad (5.44)$$

After substitution of Eq. (5.44) into (5.9), the derivative of the following quadratic form with respect to time is identified in its left side:

$$E_t(t) = \frac{1}{2} \int\limits_{\Omega} \left[\rho (u_i'^2 + T'^2 + \gamma_v T_v'^2) + \frac{\rho'^2}{\gamma M^2} \right] d\Omega.$$

The energy balance equation transformed by such a way is written as follows

$$\frac{dE_t}{dt} \equiv \Phi = -\int_{\Omega} \Bigg\{ (1+\rho')u'_i u'_j \frac{\partial U_{s,i}}{\partial x_j} + \frac{\gamma_v}{\tau}(1+\rho')(T'_v - T')^2$$

$$+\frac{\rho'}{\gamma \mathrm{M}^2}\frac{\partial \rho' u'_i}{\partial x_i} + \left((\gamma-1)(1+T') - \frac{1}{\gamma \mathrm{M}^2}\right)(1+\rho')T'\frac{\partial u'_i}{\partial x_i}$$

$$-\frac{\gamma(\gamma-1)\mathrm{M}^2\, T'}{2\mathrm{Re}}\left[\left(\frac{\partial u'_i}{\partial x_j} + \frac{\partial u'_j}{\partial x_i}\right)^2 + 2\left(\frac{\partial u'_i}{\partial x_j} + \frac{\partial u'_j}{\partial x_i}\right)\left(\frac{\partial U_{s,i}}{\partial x_j} + \frac{\partial U_{s,j}}{\partial x_i}\right)\right.$$

$$\left. + \left(\frac{\partial U_{s,i}}{\partial x_j} + \frac{\partial U_{s,j}}{\partial x_i}\right)^2 + 2\left(\alpha_1 - \frac{2}{3}\right)\left(\frac{\partial u'_i}{\partial x_i}\right)^2\right] + \frac{1}{\mathrm{Re}}\left[\left(\frac{\partial u'_i}{\partial x_j}\right)^2\right.$$

$$\left. + \left(\alpha_1 + \frac{1}{3}\right)\left(\frac{\partial u'_i}{\partial x_i}\right)^2 + \frac{\gamma}{\mathrm{Pr}}\left(\left(\frac{\partial T'}{\partial x_i}\right)^2 + \frac{20\gamma_v}{33}\left(\frac{\partial T'_v}{\partial x_i}\right)^2\right)\right]\Bigg\} d\Omega. \tag{5.45}$$

The quadratic form of E_t determines the total pulsed energy of disturbances. The last square brackets in Eq. (5.45) contain a group of positively determined terms; therefore, the right side of this equation becomes negative as the Reynolds number decreases (beginning from a certain critical value Re_{cr}). In this case one has $dE_t/dt < 0$, and all disturbances decay. The critical Reynolds number Re_{cr} corresponds to the condition of neutral disturbances $dE_t/dt = 0$ and is calculated as the minimum of the functional in the right side of the energy equation.

For further simplification of Eq. (5.45) it was assumed that the fourth-order correlations in terms of disturbances are small and can be neglected. Similar to the Sect. 5.1 (see also [19, 20]), one applies partial separation of variables in Eq. (5.45). The dependencies of velocity, density, and temperature disturbances on the periodic coordinate x_3 are presented as

$$\begin{aligned} u'_1 &= u''_1(x_1, x_2)\cos(\delta x_3), \quad u'_2 = u''_2(x_1, x_2)\cos(\delta x_3),\\ u'_3 &= u''_3(x_1, x_2)\sin(\delta x_3), \quad \rho' = \rho''(x_1, x_2)\cos(\delta x_3),\\ T' &= T''(x_1, x_2)\cos(\delta x_3), \quad T'_v = T''_v(x_1, x_2)\cos(\delta x_3). \end{aligned} \tag{5.46}$$

Using presentation (5.46), one can perform integration over the variable x_3 in the interval $[-\pi/\delta; \pi/\delta]$ in Eq. (5.45). As a result, the energy equation for the functions of the variables x_1 and x_2 takes the form

$$\frac{dE''_t}{dt} \equiv \Phi'' = -\int_S \Bigg\{ u''_1 u''_2 + \frac{1}{\mathrm{Re}}\left[\left(\frac{\partial u''_1}{\partial x_1}\right)^2 + \left(\frac{\partial u''_1}{\partial x_2}\right)^2 + \left(\frac{\partial u''_2}{\partial x_1}\right)^2\right.$$

$$+\left(\frac{\partial u_2''}{\partial x_2}\right)^2+\left(\frac{\partial u_3''}{\partial x_1}\right)^2+\left(\frac{\partial u_3''}{\partial x_2}\right)^2+\delta^2\left(u_1''^2+u_2''^2+u_3''^2\right)$$

$$+\left(\alpha_1+\frac{1}{3}\right)\left(\frac{\partial u_1''}{\partial x_1}+\frac{\partial u_2''}{\partial x_2}+\delta u_3''\right)^2+\frac{\gamma}{\Pr}\left\{\left(\frac{\partial T''}{\partial x_1}\right)^2\right.$$

$$+\left(\frac{\partial T''}{\partial x_2}\right)^2+\delta^2 T''^2+\frac{20\gamma_v}{33}\left[\left(\frac{\partial T_v''}{\partial x_1}\right)^2+\left(\frac{\partial T_v''}{\partial x_2}\right)^2+\delta^2 T_v''^2\right]\Bigg\}$$

$$-2\gamma(\gamma-1)\mathrm{M}^2 T''\left(\frac{\partial u_1''}{\partial x_2}+\frac{\partial u_2''}{\partial x_1}\right)\Bigg]+\frac{\gamma_v\left(T_v''-T''\right)^2}{\tau}$$

$$+\left(\gamma-\frac{1+\gamma\mathrm{M}^2}{\gamma\mathrm{M}^2}\right)T''\left(\frac{\partial u_1''}{\partial x_1}+\frac{\partial u_2''}{\partial x_2}+\delta u_3''\right)\Bigg\}\,dS. \tag{5.47}$$

It has been shown [12] that the operations of variation and partial integration with respect to homogeneous coordinates are permutable, and the change in the order of their performance does not alter the initial variational problem. At the same time the varied functional Φ'' in the right side becomes quadratic in terms of the disturbance functions u_i'', T'', and T_v'', and their derivatives.

5.2.2 *Asymptotics of Low Critical Reynolds Numbers*

The condition of the extremum of the functional in the right hand side of (5.47) on the set of admissible functions yields the Euler-Lagrange equations determining the generalized differential eigenvalue problem with the spectral parameter Re

$$\Delta u_1''+\left(\alpha_1+\frac{1}{3}\right)\frac{\partial D}{\partial x_1}-\gamma(\gamma-1)\mathrm{M}^2\frac{\partial T''}{\partial x_2}-\frac{\mathrm{Re}}{2}\left[u_2''-\left(\gamma-\frac{1+\gamma\mathrm{M}^2}{\gamma\mathrm{M}^2}\right)\frac{\partial T''}{\partial x_1}\right]=0,$$

$$\Delta u_2''+\left(\alpha_1+\frac{1}{3}\right)\frac{\partial D}{\partial x_2}-\gamma(\gamma-1)\mathrm{M}^2\frac{\partial T''}{\partial x_1}-\frac{\mathrm{Re}}{2}\left[u_1''-\left(\gamma-\frac{1+\gamma\mathrm{M}^2}{\gamma\mathrm{M}^2}\right)\frac{\partial T''}{\partial x_2}\right]=0,$$

$$\Delta u_3''-\delta\left(\alpha_1+\frac{1}{3}\right)D-\frac{\mathrm{Re}}{2}\left(\gamma-\frac{1+\gamma\mathrm{M}^2}{\gamma\mathrm{M}^2}\right)\delta T''=0,$$

$$\frac{\gamma}{\Pr}\Delta T''+\gamma(\gamma-1)\mathrm{M}^2\left(\frac{\partial u_1''}{\partial x_2}+\frac{\partial u_2''}{\partial x_1}\right)-\frac{\mathrm{Re}}{2}\left[2\gamma_v\frac{T''-T_v''}{\tau}+\left(\gamma-\frac{1+\gamma\mathrm{M}^2}{\gamma\mathrm{M}^2}\right)D\right]=0,$$

$$\frac{\gamma}{\Pr}\Delta T_v''-\frac{\mathrm{Re}}{2}\frac{33\left(T_v''-T''\right)}{10\,\tau}=0. \tag{5.48}$$

Here, one has

$$D = \frac{\partial u''_1}{\partial x_1} + \frac{\partial u''_2}{\partial x_2} + \delta u''_3, \quad \Delta = \frac{\partial^2}{\partial x_1^2} + \frac{\partial^2}{\partial x_2^2} - \delta^2,$$

the amplitude functions u''_i, ρ'', T'', and T''_v satisfy periodic boundary conditions at $x_1 = \pm\pi/\alpha$ and take zero values on the impermeable boundaries $x_2 = \pm 1$.

The solution of system (5.48) $\boldsymbol{\varphi}'' = (u''_1, u''_2, u''_3, T'', T''_v)$ is searched for in the form

$$\boldsymbol{\varphi}''(x_1, x_2) = \boldsymbol{\varphi}(x_2)\exp(i\alpha x_1),$$

where $\boldsymbol{\varphi} = (u, v, w, \theta, \theta_v)$, α is the projection of the real wave vector onto the coordinate axis x_1, and i is the imaginary unit. Then the spectral problem reduces to a system of ordinary differential equations for the amplitudes of fluctuations of u, v, w, θ, and θ_v

$$u'' + i\alpha\left(\alpha_1 + \frac{1}{3}\right)v' - \left[\alpha^2\left(\alpha_1 + \frac{4}{3}\right) + \delta^2\right]u$$

$$+ i\alpha\delta\left(\alpha_1 + \frac{1}{3}\right)w - \gamma(\gamma-1)\mathrm{M}^2\theta' = \frac{\mathrm{Re}}{2}\left[v + \left(1 - \gamma + \frac{1}{\gamma\mathrm{M}^2}\right)i\alpha\theta\right],$$

$$\left(\alpha_1 + \frac{4}{3}\right)v'' + i\beta\left(\alpha_1 + \frac{1}{3}\right)u' + \delta\left(\alpha_1 + \frac{1}{3}\right)w' - k^2 v$$

$$-\gamma(\gamma-1)\mathrm{M}^2\, i\beta\theta = \frac{\mathrm{Re}}{2}\left[u + \left(1 - \gamma + \frac{1}{\gamma\mathrm{M}^2}\right)\theta'\right],$$

$$w'' - \delta\left(\alpha_1 + \frac{1}{3}\right)v' - i\beta\delta\left(\alpha_1 + \frac{1}{3}\right)u - \left[\delta^2\left(\alpha_1 + \frac{4}{3}\right) + \beta^2\right]w$$

$$= -\frac{\mathrm{Re}}{2}\left(1 - \gamma + \frac{1}{\gamma\mathrm{M}^2}\right)\delta\theta,$$

$$\frac{\gamma}{\mathrm{Pr}}\left(\theta'' - k^2\theta\right) + \gamma(\gamma-1)\mathrm{M}^2\left(u' + i\beta v\right)$$

$$= \frac{\mathrm{Re}}{2}\left[\frac{2\gamma_v}{\tau}(\theta - \theta_v) - \left(1 - \gamma + \frac{1}{\gamma\mathrm{M}^2}\right)\left(v' + i\beta u + \delta w\right)\right],$$

$$\frac{\gamma}{\mathrm{Pr}}\left(\theta''_v - k^2\theta_v\right) = -\frac{\mathrm{Re}}{2}\,\frac{33\,(\theta - \theta_v)}{10\,\tau}, \quad k^2 = \alpha^2 + \delta^2. \tag{5.49}$$

Hereinafter, the primes at the unknown functions mean their derivatives of the corresponding order with respect to the variable x_2. Problem (5.49) is closed by the boundary conditions

$$u\big|_{x_2=\pm 1} = v\big|_{x_2=\pm 1} = w\big|_{x_2=\pm 1} = \theta\big|_{x_2=\pm 1} = \theta_v\big|_{x_2=\pm 1} = 0. \tag{5.50}$$

The spectral problem (5.49), (5.50) has the following properties.

(i) The spectrum of the eigenvalues Re of problem (5.49), (5.50) is real. This property follows from the energy identity for system (5.49), (5.50), which is obtained by multiplying Eq. (5.49) by the complex-conjugate functions u^*, v^*, w^*, θ^*, and θ_v^*, their summation, and integration over the interval $x_2 \in [-1;\ 1]$. In view of the homogeneous boundary conditions (5.50), we obtain the expression

$$\frac{\mathrm{Re}}{2}\int_{-1}^{1}\Bigg\{u^* v + u v^* + \frac{2\gamma_v}{\tau}\,|\,\theta - \theta_v|^2$$

$$+\left(1-\gamma+\frac{1}{\gamma\,\mathrm{M}^2}\right)\Bigg[2\alpha\left(\theta_r u_i - \theta_i u_r\right) - \delta\left(\theta^* w + \theta w^*\right) + v_r\frac{d\theta_r}{dx_2}$$

$$+v_i\frac{d\theta_i}{dx_2} - \theta_r\frac{dv_r}{dx_2} - \theta_i\frac{dv_i}{dx_2}\Bigg]\Bigg\}dx_2 + \int_{-1}^{1}\Bigg\{\left|\frac{du}{dx_2}\right|^2 + \left(\alpha_1 + \frac{4}{3}\right)\left|\frac{dv}{dx_2}\right|^2$$

$$+\left|\frac{dw}{dx_2}\right|^2 + \frac{\gamma}{\mathrm{Pr}}\left[\left|\frac{d\theta}{dx_2}\right|^2 + k^2|\,\theta|^2 + \frac{20\gamma_v}{33}\left(\left|\frac{d\theta_v}{dx_2}\right|^2 + k^2|\,\theta_v|^2\right)\right]$$

$$+\left(\alpha^2\left(\alpha_1+\frac{1}{3}\right)+k^2\right)|\,u|^2 + k^2|\,v|^2 + \left(\delta^2\left(\alpha_1+\frac{1}{3}\right)+k^2\right)|\,w|^2$$

$$+\left(\alpha_1+\frac{1}{3}\right)\Bigg[\alpha\left(u_r\frac{dv_i}{dx_2} + v_r\frac{du_i}{dx_2} - u_i\frac{dv_r}{dx_2} - v_i\frac{du_r}{dx_2}\right)$$

$$-\delta\left(v_r\frac{dw_r}{dx_2} + v_i\frac{dw_i}{dx_2} - w_r\frac{dv_r}{dx_2} - w_i\frac{dv_i}{dx_2}\right) + 2\alpha\delta\left(u_r w_i - u_i w_r\right)\Bigg]$$

$$-\gamma\left(\gamma-1\right)\mathrm{M}^2\Bigg[\delta\left(\theta_r\frac{du_r}{dx_2} + \theta_i\frac{du_i}{dx_2} - u_r\frac{d\theta_r}{dx_2} - u_i\frac{d\theta_i}{dx_2}\right)$$

$$-2\alpha\left(\theta_r v_i - \theta_i v_r\right)\Bigg]\Bigg\}dx_2 = 0. \tag{5.51}$$

Here the subscripts r and i indicate the real and imaginary parts of the corresponding complex-valued functions, respectively. The fact that the spectral parameter Re is real is determined by the fact that all terms of identity (5.51) are real. At the same time the quadratic form determined by the energy identity (5.51) is not positively determined. This means that the eigenvalues can also be negative; therefore, we have to search for the minimum absolute value, min |Re|, in our calculations.

(*ii*) The spectrum of the eigenvalues $\mathrm{Re}(\alpha, \delta)$ is symmetric with respect to the axes $\alpha = 0$ and $\delta = 0$ on the plane of wavenumbers (α, δ). Indeed, it follows from the equations of system (5.49) that each eigenvalue $\mathrm{Re}(\alpha, \delta)$ with the eigenfunctions u, v, w, θ, and θ_v corresponds to an equal eigenvalue $\mathrm{Re}(-\alpha, \delta)$ with the eigenfunctions u^*, v^*, w^*, θ^*, and θ_v^*. The same eigenvalue $\mathrm{Re}(\alpha, \delta)$ also corresponds to a pair of wavenumbers $(\alpha, -\delta)$ and to a set of eigenfunctions u, v, $-w$, θ, and θ_v.

(*iii*) The eigenvalues for particular disturbance modes have asymptotic solutions. It was considered two limiting cases in which we can pass from the spectral problem (5.49), (5.50) to the scalar polynomial spectral bundle of ordinary differential operators. To find the asymptotic solution of the eigenvalues of this bundle, one can use the method developed by Tamarkin [21] who considered a spectral bundle of operators of the form

$$y^{(n)} + p_1 y^{(n-1)} + \cdots + p_n y = 0, \tag{5.52}$$

whose coefficients are polynomials of the spectral parameter λ:

$$p_i(x_2, \lambda) = \lambda^i \sum_{j=0}^{i} \lambda^{-j} p_{ij}(x_2).$$

Equation (5.52) is supplemented with a system of homogeneous boundary conditions, which can also contain polynomials in terms of λ with a power not higher than n. To obtain the asymptotic solution for eigenvalues outside a certain circle on the complex plane of the spectral parameter, it was considered the following simplified equation instead of Eq. (5.52):

$$y^{(n)} + \lambda p_{10} y^{(n-1)} + \cdots + \lambda^k p_{k0} y^{(n-k)} + \cdots + \lambda^n p_{n0} y = 0. \tag{5.53}$$

When replacing Eq. (5.52) by (5.53), one has to leave in the latter equation only those terms in which the sum of the higher power of the spectral parameter and the order of the derivative coincides with the order of Eq. (5.52).

Substitution of the expression $y \sim \exp(\mu x_2)$ into Eq. (5.53) yields the characteristic equation

$$\mu^n + \lambda p_{10} \mu^{n-1} + \cdots + \lambda^k p_{k0} \mu^{n-k} + \cdots + \lambda^n p_{n0} = 0,$$

which is reduced to the following form by the self-similar replacement of the variable $\mu = \lambda z$:

$$z^n + p_{10} z^{n-1} + \cdots + p_{k0} z^{n-k} + \cdots + p_{n0} = 0. \tag{5.54}$$

After the roots of Eq. (5.54) are found, the system of the solutions of Eq. (5.53) $y_i = \exp(\lambda z_i x_2)$ $(i = 1, \ldots, n)$ is substituted into the boundary conditions, and the asymptotic eigenvalues of the initial spectral problem are found.

Using the above-described approach two limiting modes of disturbances were studied.

Constant Mode $\alpha = \delta = 0$. In this case system (5.49) takes the form

$$u'' - a\theta' - \lambda v = 0, \quad bv'' + \lambda(c\theta' - u) = 0, \quad w'' = 0,$$

$$d\theta'' + au' - \lambda(cv' + f\vartheta) = 0, \quad g\theta_v'' + \lambda f \vartheta = 0.$$

Here a, b, c, d, f, and g are the coefficients:

$$a = \gamma(\gamma - 1)\mathrm{M}^2, \quad b = \alpha_1 + 4/3, \quad c = \gamma - 1 - 1/(\gamma \mathrm{M}^2),$$

$$d = \gamma/\mathrm{Pr}, \quad f = 2\gamma\gamma_v/\tau, \quad g = (20/33)\gamma_v\gamma^2/\mathrm{Pr}. \tag{5.55}$$

Additionally, the notations $\lambda = \mathrm{Re}/2$ and $\vartheta = \theta - \theta_v$ are introduced.

The equation for the function w is integrated separately and has the general solution $w = c_1 x_2 + c_2$, which is identically equal to zero by virtue of the homogeneous boundary conditions (5.50), which reduces the order of the system by two unities. The sequential elimination of unknowns for this system yields a scalar spectral problem for the function $\vartheta(x_2)$. To reduce the problem to the problem for the spectral bundle of the form (5.53), all terms that did not satisfy the necessary requirement were eliminated at intermediate stages. As a result, one obtains the equation

$$\vartheta^{(6)} + \lambda r \vartheta^{(4)} + \lambda^2 s \vartheta^{(2)} + \lambda^3 t \vartheta = 0, \tag{5.56}$$

where r, s, and t are the coefficients:

$$r = -\frac{f(d+g)}{dg}, \quad s = \frac{c^2 - ac - d}{bd},$$

$$t = \frac{f}{bd}\left[\frac{1}{g}\left(\frac{ac}{d} + 1\right)^2 (d+g) + \left(\frac{ac}{d} + 1\right)\frac{a^2}{d} - c^2(d+g) + \frac{ac}{d}\right].$$

By replacing the variable $y^{(k)} = \vartheta^{(2k)}$ $(k = 0, 1, 2, 3)$, we reduce Eq. (5.53) to an equation of the form (5.53):

$$y''' + \lambda r y'' + \lambda^2 s y' + \lambda^3 t y = 0. \tag{5.57}$$

In this case the following complete cubic equation is the characteristic equation corresponding to Eq. (5.54):

$$z^3 + rz^2 + sz + t = 0. \tag{5.58}$$

To analyze its roots a standard approach [13] is used. Using the linear replacement of the variable $z = x - r/3$, one passes to the reduced cubic equation

$$x^3 + px + q = 0,$$

where

$$p = (3s - r^2)/3, \quad q = 2r^3/27 - sr/3 + t.$$

To study the roots of this equation its discriminant

$$D = (p/3)^3 + (q/2)^2$$

is considered. The estimates show that the inequalities $p < 0$, $q < 0$, and

$$|(p/3)^3| \ll (q/2)^2 \tag{5.59}$$

are valid in the entire range of the values of the coefficients in the initial equation (5.56). Therefore, the discriminant D is greater than zero, and Eq. (5.58) (or Eq. (5.57)) has one real root and two complex-conjugate roots, which are determined via auxiliary functions u and v by the Cardano formulas [13]

$$x_1 = u + v, \quad x_{2,3} = \frac{u+v}{2} \pm \frac{u-v}{2} i\sqrt{3}.$$

Here $u = (-q/2 + \sqrt{D})^{1/3}$ and $v = (-q/2 - \sqrt{D})^{1/3}$. By virtue of Eq. (5.59), to obtain the asymptotic estimate, one can take $\sqrt{D} \simeq |q/2|$. Then, it is obtained $u \simeq |q|^{1/3}$ and $v \simeq 0$. As a result, one has

$$x_1 = |q|^{1/3}, \quad x_{2,3} = -|q|^{1/3}\left(1 \mp i\sqrt{3}\right)/2.$$

Thus, the roots of Eq. (5.58) have the form

$$z_1 = |q|^{1/3} + \frac{\Pr}{2\tau\gamma}\left(1 + \frac{20}{33}\gamma_v\gamma\right),$$

$$z_{2,\,3} = -\frac{\Pr}{2\tau\gamma}\left(1 + \frac{20}{33}\gamma_v\gamma\right)|q|^{1/3}\,\frac{1 \mp i\sqrt{3}}{2}.$$

Finally, the characteristic equation corresponding to the initial equation (5.56) has a pair of real roots

$$\delta_{1,2} = \pm\sqrt{\left|\frac{\mathrm{Re}}{2}\right|\left[|q|^{1/3} + \frac{\mathrm{Pr}}{2\tau\gamma}\left(1 + \frac{20}{33}\gamma_v\gamma\right)\right]} \equiv \pm\delta$$

and two pairs of complex-conjugate roots

$$\delta_{3,4} = \pm\sqrt{|\mu|}\exp\left(\frac{i}{2}\arg\mu\right), \quad \delta_{5,6} = \pm\sqrt{|\mu|}\exp\left(-\frac{i}{2}\arg\mu\right).$$

In this case one has

$$|\mu| = \frac{|\mathrm{Re}|}{4}\sqrt{\left[|q|^{1/3} - \frac{\mathrm{Pr}}{\tau\gamma}\left(1 + \frac{20}{33}\gamma_v\gamma\right)\right]^2 + 3\,|q|^{2/3}},$$

$$\arg\mu = \arctan\frac{|q|^{1/3}\sqrt{3}}{|q|^{1/3} - \mathrm{Pr}\left[1 + (20/33)\gamma_v\gamma\right]/(\tau\gamma)}.$$

To avoid complication of the boundary conditions for Eq. (5.56), one can confine to paired considerations of the solutions under the boundary conditions $\vartheta(\pm 1) = 0$. For the pair of the real roots $\delta_{1,2}$, the solution is sought in the form

$$\vartheta(x_2) = C_1\exp(\delta x_2) + C_2\exp(-\delta x_2).$$

Substituting the solution into the boundary conditions, one obtains a homogeneous system of equations for arbitrary constants:

$$C_1\exp(\delta) + C_2\exp(-\delta) = 0,$$

$$C_1\exp(-\delta) + C_2\exp(\delta) = 0.$$

At $\delta \neq 0$, this system has only the trivial solution $C_1 = C_2 = 0$.

As the coefficients of Eq. (5.56) are real, the analysis of the complex-conjugate roots can be limited to separate consideration of the real and imaginary parts of the solution

$$\vartheta(x_2) = C_3\exp[\sqrt{|\mu|}\,(\cos\varphi + i\sin\varphi)x_2]$$

$$+ C_4\exp[-\sqrt{|\mu|}\,(\cos\varphi + i\sin\varphi)x_2], \tag{5.60}$$

where $\varphi = (1/2)\arg\mu$.

Substitution of the real part of Eq. (5.60) into the boundary conditions yields the expressions

$$[C_3\exp(\sqrt{|\mu|}\,\cos(\sqrt{|\mu|}\,\sin\varphi))$$

$$+C_4\exp(-\sqrt{|\mu|}\,\cos\varphi)]\cos(\sqrt{|\mu|}\,\sin\varphi) = 0,$$

$$[C_3 \exp(-\sqrt{|\mu|}\cos(\sqrt{|\mu|}\sin\varphi))$$

$$+C_4 \exp(\sqrt{|\mu|}\cos\varphi)]\cos(\sqrt{|\mu|}\sin\varphi) = 0.$$

For the system to have nontrivial solutions, one has to satisfy the condition

$$\cos(\sqrt{|\mu|}\sin\varphi) = 0,$$

which yields the equations for the eigenvalues Re:

$$\sqrt{|\mu|}\sin\varphi = (2n-1)\pi/2, \quad n = 1,\ 2,\ \ldots .$$

Therefore, one has

$$|\mathrm{Re}| = \frac{\pi^2(2n-1)^2}{\sin^2\varphi}\left\{\left[|q|^{1/3} - \frac{\mathrm{Pr}}{\tau\gamma}\left(1 + \frac{20}{33}\gamma_v\gamma\right)\right]^2 + 3|q|^{2/3}\right\}^{-1/2},$$

$$n = 1,\ 2,\ \ldots . \tag{5.61}$$

Substitution of the imaginary part of Eq. (5.60) into the boundary conditions yields the system of equations

$$[C_3 \exp(\sqrt{|\mu|}\cos(\sqrt{|\mu|}\sin\varphi))$$

$$-C_4 \exp(-\sqrt{|\mu|}\cos\varphi)]\sin(\sqrt{|\mu|}\sin\varphi) = 0,$$

$$[C_3 \exp(-\sqrt{|\mu|}\cos(\sqrt{|\mu|}\sin\varphi))$$

$$-C_4 \exp(\sqrt{|\mu|}\cos\varphi)]\sin(\sqrt{|\mu|}\sin\varphi) = 0.$$

In this case, the condition of existence of nontrivial solutions is determined by the equation

$$\sin(\sqrt{|\mu|}\sin\varphi) = 0,$$

whose roots are $\sqrt{|\mu|}\sin\varphi = \pi n$ $(n = 0, 1, \ldots)$. These roots correspond to the set of eigenvalues

$$|\mathrm{Re}| = \frac{\pi^2 n^2}{\sin^2\varphi}\left\{\left[|q|^{1/3} - \frac{\mathrm{Pr}}{\tau\gamma}\left(1 + \frac{20}{33}\gamma_v\gamma\right)\right]^2 + 3|q|^{2/3}\right\}^{-1/2},$$

$$n = 1,\ 2,\ \ldots . \tag{5.62}$$

At $n = 1$ the sets of eigenvalues (5.61) and (5.62) have an identical minimum corresponding to the critical Reynolds number $\mathrm{Re}_{\mathrm{cr}}$. Expressing the value of $|q|$ in Eq. (5.62) via coefficients (5.55) of the initial system, one obtains the asymptotic formula

$$\mathrm{Re}_{cr} = \pi^2 \left(\frac{a}{\tau^3} + \frac{b}{\tau \mathrm{M}^4(\alpha_1 + 4/3)} \right)^{-1/3}, \tag{5.63}$$

where a and b are certain positive constants of the order $O(1)$.

Streamwise Long-wave mode $\alpha \ll 1$, $\delta = 0$. In this case neglecting terms of the order $O(\alpha^2)$ in system (5.49), one can obtain

$$u'' + (\alpha_1 + 1/3)i\alpha v' - a\theta' + \lambda(i\alpha\theta' - v) = 0,$$

$$bv'' + (\alpha_1 + 1/3)i\alpha u' - ai\alpha\theta + \lambda(c\theta' - u) = 0,$$

$$w'' = 0, \quad d\theta'' + a(u' + i\alpha v) - \lambda[c(v' + i\alpha u) + f\theta] = 0, \quad g\theta_v'' + \lambda f\theta = 0.$$

Here the coefficients a, b, c, d, f, and g are determined by relations (5.55). The equation for the function w is integrated separately and has only the zero solution $w = 0$ under the homogeneous boundary conditions (5.50).

The reduced system transforms as follows. In accordance with the approach [22] to the asymptotic analysis of the spectral problems for systems of ordinary differential equations, all lower-order terms that do not contain the spectral parameter λ are eliminated from the above-given system. Further transformations similar to the transformations for the constant mode yield an equation of the form (5.57) whose coefficients are determined by the relations

$$r = -f\frac{d+g}{dg}, \quad s = \frac{c^2}{d}\Big(\frac{1}{b} - \alpha^2\Big), \quad t = \frac{c^2}{d} f\Big(\frac{\alpha^2}{g} - \frac{1}{b}\Big).$$

These relations contain terms with α^2, which can be neglected as in the initial system. The analysis of the resultant spectral problem coincides with the analysis performed above for the constant mode. As a result an asymptotic formula similar to Eq. (5.63) for the critical Reynolds number was obtained, where the coefficients a and b have inessential differences.

5.2.3 Numerical Calculation of Low Critical Reynolds Numbers

The spectral problem (5.49), (5.50) with arbitrary finite values of wave numbers was solved numerically with the use of the Matlab software system. The method of collocations [16] previously used in Sects. 3.2, 5.1 (see also [19]) was applied. The problem was written in the matrix presentation as

$$A\,\frac{d^2\varphi}{d\,x_2^2}+B_1\,\frac{d\,\varphi}{d\,x_2}+C_1\,\varphi=\mathrm{Re}\left(B_2\,\frac{d\,\varphi}{d\,x_2}+C_2\,\varphi\right),\quad \varphi\big|_{x_2=\pm1}=0. \tag{5.64}$$

Here A, B_1, C_1, B_2, and C_2 are matrices of size 5×5:

$$A=\begin{pmatrix} 1 & 0 & 0\,0\,0\\ 0 & \alpha_1+\frac{4}{3} & 0\,0\,0\\ 0 & 0 & 1\,0\,0\\ 0 & 0 & 0\,a\,0\\ 0 & 0 & 0\,0\,a \end{pmatrix},$$

$$B_1=\begin{pmatrix} 0 & i\alpha\left(\alpha_1+\frac{1}{3}\right) & 0 & -b & 0\\ i\alpha\left(\alpha_1+\frac{1}{3}\right) & 0 & \delta\left(\alpha_1+\frac{1}{3}\right) & 0 & 0\\ 0 & -\delta\left(\alpha_1+\frac{1}{3}\right) & 0 & 0 & 0\\ b & 0 & 0 & 0 & 0\\ 0 & 0 & 0 & 0 & 0 \end{pmatrix},$$

$$C_1=\begin{pmatrix} -d_1 & 0 & i\alpha\delta\left(\alpha_1+\frac{1}{3}\right) & 0 & 0\\ 0 & -k^2 & 0 & -i\alpha b & 0\\ -i\alpha\delta\left(\alpha_1+\frac{1}{3}\right) & 0 & -d_2 & 0 & 0\\ 0 & i\alpha b & 0 & -ak^2 & 0\\ 0 & 0 & 0 & 0 & -ak^2 \end{pmatrix},$$

$$B_2=\begin{pmatrix} 0 & 0 & 0 & 0 & 0\\ 0 & 0 & 0 & c/2 & 0\\ 0 & 0 & 0 & 0 & 0\\ 0 & -c/2 & 0 & 0 & 0\\ 0 & 0 & 0 & 0 & 0 \end{pmatrix},$$

$$C_2=\begin{pmatrix} 0 & 1/2 & 0 & i\alpha c/2 & 0\\ 1/2 & 0 & 0 & 0 & 0\\ 0 & 0 & 0 & -\delta c/2 & 0\\ -i\alpha c/2 & 0 & -\delta c/2 & \gamma_v/\tau & -\gamma_v/\tau\\ 0 & 0 & 0 & -33/(20\,\tau) & 33/(20\,\tau) \end{pmatrix},$$

$$a=\frac{\gamma}{\mathrm{Pr}},\quad b=\gamma\,(\gamma-1)\,\mathrm{M}^2,\quad c=1-\gamma+1/\left(\gamma\,\mathrm{M}^2\right),$$

$$d_1 = \alpha^2\left(\alpha_1 + \frac{1}{3}\right) + k^2, \quad d_2 = \delta^2\left(\alpha_1 + \frac{1}{3}\right) + k^2, \quad k^2 = \alpha^2 + \delta^2.$$

The first-order differential operators involved in the spectral problem are approximated on this grid by a matrix of the collocation derivatives (3.31). The second-order differential operators are approximated by the superposition (5.35). The thus-constructed discrete approximation allows us to reduce problem (5.64) to a generalized eigenvalue problem for linear matrix bundle of the form (5.36) with respect to the spectral parameter $\lambda = \text{Re}$. Where the vector of unknowns $\boldsymbol{q}$ of size $5(N+1)$ consists of the values of the eigenfunctions in the collocation nodes:

$$\boldsymbol{q} = (u_0, u_1, \ldots, u_N, v_0, v_1, \ldots, v_N, w_0, w_1, \ldots, w_N, \\ \theta_0, \theta_1, \ldots, \theta_N, \theta_{v,0}, \theta_{v,1}, \ldots, \theta_{v,N}),$$

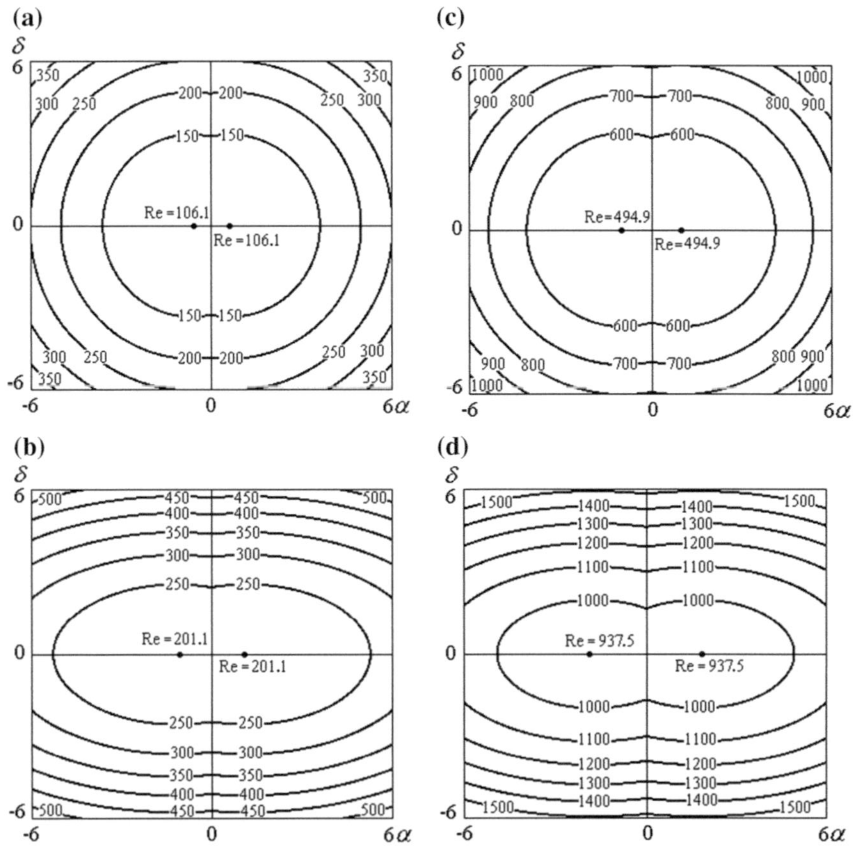

Fig. 5.4 Level lines of the surfaces $\text{Re}(\alpha, \delta)$ for $\alpha_1 = 0$ and $\tau = 2$. **a**, **b** are $\text{M} = 3$. **c**, **d** are $\text{M} = 5$. **a**, **c** are $\gamma_v = 0$. **b**, **d** are 0.667. The points on the line $\delta = 0$ are the critical values of the Reynolds number for corresponding regime

the matrices G and F of size $5(N+1) \times 5(N+1)$ are calculated by a special Matlab procedure by the formulas

$$G = A \otimes D_N^2 + B_1 \otimes D_N^1 + C_1 \otimes I_N,$$

$$F = B_2 \otimes D_N^1 + C_2 \otimes I_N.$$

The homogeneous boundary conditions for Eq. (5.64) are taken into account via the conditions (5.37).

By using this procedure for fixed values of the Mach number M, bulk viscosity α_1, degree of nonequilibrium of the vibrational energy γ_v, vibrational relaxation time τ, and each pair of wavenumbers (α, δ), a set of $N+1$ eigenvalues was obtained, which include the Reynolds number with the minimum absolute value $\mathrm{Re}(\alpha, \delta) = |\lambda_{\min}(\alpha, \delta)|$. The value of the critical Reynolds number $\mathrm{Re}_{\mathrm{cr}}$ for the chosen values of M, α_1, γ_v, and τ was taken to be the minimum value of Re in the entire range of wavenumbers $\mathrm{Re}(\alpha, \delta)$: $\mathrm{Re}_{cr} = \min\limits_{(\alpha, \delta)} [\mathrm{Re}(\alpha, \delta)]$. After that, the eigenfunctions u, v, w, θ, and θ_v corresponding to Re_{cr} were calculated.

The spectra of the eigenvalues $\lambda(\alpha, \delta, \mathrm{M}, \alpha_1, \gamma_v, \tau)$ were calculated in the range of wavenumbers $\alpha = 0$–5 and $\delta = 0$–5 with the following values of the parameters: $\gamma_v = 0, 0.667, \tau = 1$–$4, \alpha_1 = 0$–$2, \mathrm{M} = 0.1$–$5, \mathrm{Pr} = 3/4, \gamma = 7/5$. The step of wave number variation was chosen to be $h_\alpha = h_\delta = 0.1$. In most calculations the number of collocation nodes in the interval $[-1;\ 1]$ was taken to be $N+1 = 50$. To check the calculation accuracy test calculations were performed in the range of wavenumbers $\alpha = -5$–5, $\delta = -5$–5 by virtue of spectrum symmetry and using variation of the number of the collocation nodes in the limits $N+1 = 32$–100.

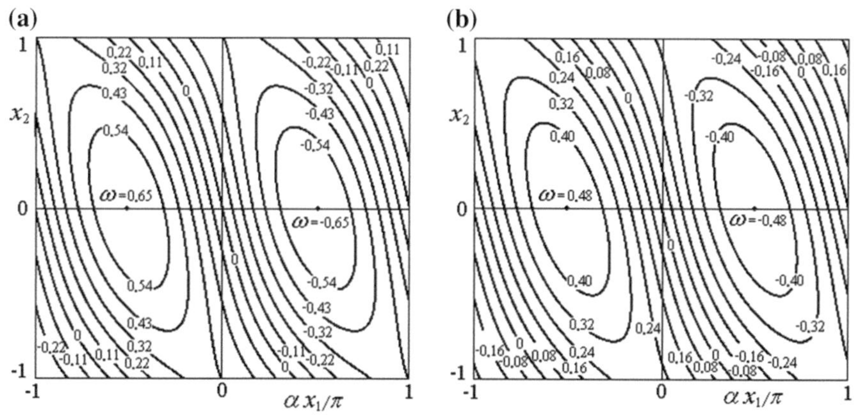

Fig. 5.5 Isolines of vorticity of the critical disturbances $\omega(x_1, x_2)$ at $\mathrm{M} = 3$, $\alpha_1 = 0$, and $\tau = 2$. **a** is $\gamma_v = 0$, $\mathrm{Re}_{cr} = 106.1$. **b** is $\gamma_v = 0.667$, $\mathrm{Re}_{cr} = 201.1$. The points on the line $x_2 = 0$ are the maximum and minimum values of ω for corresponding regime

As in the case of moderate excitation considered in Sect. 5.1 (see also [19]) the disturbances of the streamwise mode are the most "dangerous" ones. In view of periodicity of the solution along the streamwise coordinate x_1 these disturbances are pairs of two-dimensional vortices rotating in opposite directions with their axes being transverse to the carrier flow. The vorticity distribution in these vortices is calculated by the formula

$$\omega(x_1, x_2) = -\left(\alpha v_i + \frac{du_r}{dx_2}\right)\cos\alpha x_1 - \left(\alpha v_r - \frac{du_i}{dx_2}\right)\sin\alpha x_1.$$

Here $u_r(x_2)$, $u_i(x_2)$, $v_r(x_2)$, and $v_i(x_2)$ are the real and imaginary parts of the eigenfunctions u and v.

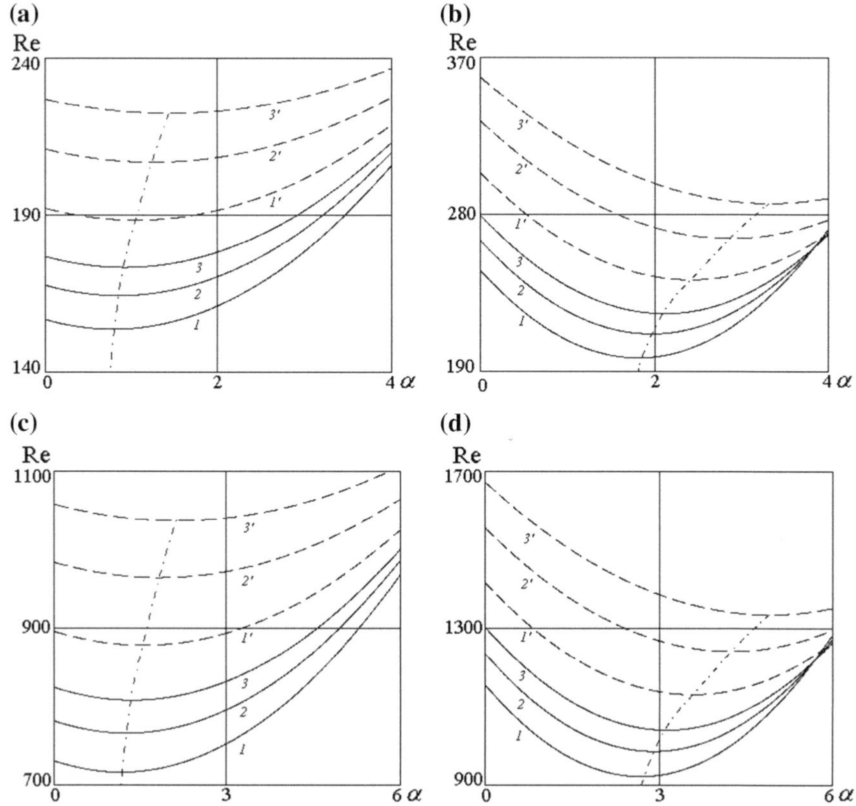

Fig. 5.6 Dependencies Re(α) (neutral stability curves) for the streamwise modes of disturbances for $\gamma_v = 0.250$ (*1* and *1'*), 0.429 (*2* and *2'*), and 0.667 (*3* and *3'*). **a**, **b** are M = 3. **c**, **d** are M = 5. **a**, **c** are $\alpha_1 = 0$. **b**, **d** are $\alpha_1 = 2$. The *solid* and *dashed curves* correspond to $\tau = 1$ and 3, respectively. The *dot-and-dashed curves* show the dependence of the critical Reynolds number Re_{cr} on the wavenumber α

The calculations show that the minimum absolute eigenvalues

$$\mathrm{Re}(\alpha, \delta) = |\lambda_{\min}(\alpha, \delta)|$$

for all considered degrees of nonequilibrium of vibrational energy γ_v, vibrational relaxation time τ, bulk viscosity α_1, and Mach number M are reached on the axis $\alpha \neq 0$ (at $\delta = 0$) in the plane of wavenumbers (α, δ). The level lines of $\mathrm{Re}(\alpha, \delta)$ are shown in Fig. 5.4.

Figure 5.5 shows the vorticity isolines $\omega(x_1, x_2)$ for different critical Reynolds numbers $\mathrm{Re}_{cr}(\alpha, \alpha_1, \gamma_v, \tau, \mathrm{M})$ and the velocity disturbance amplitudes reaching 10% of the absolute value of the carrier flow on the impermeable boundaries.

The dependence of the Reynolds number for the streamwise modes of disturbances on the wavenumber α is shown in Fig. 5.6.These dependencies one can consider as neutral stability curves (see Fig. 5.2)The dot-and-dashed curves connect the absolute minimums on the curves $\mathrm{Re}(\alpha)$ parameterized in terms of γ_v and τ, which allow to trace the evolution of Re_{cr}.

Figure 5.7 shows Re_{cr} as a function of the nonequilibrium degree γ_v. It is seen from Figs. 5.4, 5.6, and 5.7 that the critical Reynolds numbers Re_{cr} and the corresponding wavenumbers α increase with increasing values of the parameters α_1, γ_v, τ, and M. Thus, the calculation results show that the above-derived asymptotic expressions (5.63) provide a qualitatively correct description of the dependence of Re_{cr} on flow parameters.

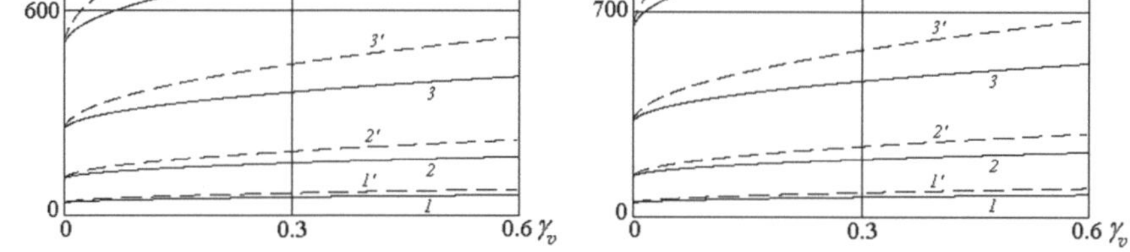

Fig. 5.7 Critical Reynolds number Re_{cr} versus the degree of nonequilibrium of the vibrational mode γ_v at M = 2 (*1* and *1′*), 3 (*2* and *2′*), 4 (*3* and *3′*), and 5 (*4* and *4′*). **a** is $\alpha_1 = 0$. **b** is $\alpha_1 = 2$. The *solid* and *dashed curves* correspond to $\tau = 1$ and 3, respectively

Table 5.2 Critical values of the Reynolds number Re_{cr} (α_1, γ_v, τ, M)

M	$\tau = 1$			$\tau = 4$		
	$\gamma_v = 0, 111$	0, 250	0, 667	0, 111	0, 250	0, 667
$\alpha_1 = 0$						
0, 5	15, 57	17, 12	19, 31	19, 31	22, 41	26, 79
0, 8	16, 73	18, 40	20, 75	20, 75	24, 08	28, 79
2, 0	48, 22	53, 02	59, 82	59, 82	69, 42	82, 98
3, 0	139, 70	153, 62	173, 28	173, 28	201, 08	240, 41
5, 0	651, 37	716, 19	807, 86	807, 86	937, 51	1120, 85
$\alpha_1 = 2$						
0, 5	20, 04	22, 04	24, 86	24, 86	28, 85	34, 49
0, 8	21, 54	23, 68	26, 71	26, 71	30, 99	37, 06
2, 0	62, 07	68, 25	76, 99	76, 99	89, 34	106, 81
3, 0	179, 83	197, 72	223, 03	223, 03	258, 82	309, 44
5, 0	838, 40	921, 84	1039, 83	1039, 83	1206, 70	1442, 69

Table 5.3 Wavenumbers α corresponding to the critical Reynolds numbers Re_{cr} (α_1, γ_v, τ, M)

M	$\tau = 1$			$\tau = 4$		
	$\gamma_v = 0, 111$	0, 250	0, 667	0, 111	0, 250	0, 667
$\alpha_1 = 0$						
0, 5	0, 488	0, 533	0, 621	0, 621	0, 799	1, 154
0, 8	0, 554	0, 604	0, 705	0, 704	0, 906	1, 309
2, 0	0, 672	0, 733	0, 855	0, 855	1, 099	1, 587
3, 0	0, 718	0, 784	0, 914	0, 914	1, 175	1, 698
5, 0	1, 063	1, 160	1, 353	1, 353	1, 739	2, 513
$\alpha_1 = 2$						
0, 5	1, 125	1, 228	1, 432	1, 432	1, 842	2, 660
0, 8	1, 277	1, 393	1, 625	1, 625	2, 089	3, 017
2, 0	1, 548	1, 689	1, 970	1, 970	2, 533	3, 658
3, 0	1, 656	1, 806	2, 107	2, 107	2, 709	3, 913
5, 0	2, 450	2, 673	3, 118	3, 119	4, 009	5, 791

The critical Reynolds numbers $\mathrm{Re}_{cr}(\alpha_1, \gamma_v, \tau, \mathrm{M})$ and the corresponding wavenumbers α are listed in Tables 5.2 and 5.3, respectively. In the considered range of the problem parameters, the Reynolds number Re_{cr} in the limit increases approximately by a factor of 2.5. The effect of each particular parameter on Re_{cr} with fixed values of the remaining parameters is appreciably different from the effect of other parameters. It should be noted that the most pronounced effect on Re_{cr} is exerted by the degree of nonequilibrium of the vibrational mode γ_v. The estimates obtained in [23] show that excitation of the vibrational mode γ_v in the considered range can be easily ensured by a laser. Therefore, laser pumping of vibrational modes can be considered as an effective method of controlling molecular gas flows.

References

1. Nerushev, A., Novopashin, S.: Rotational relaxation and transition to turbulence. Phys. Lett. A **232**, 243–245 (1997)
2. Zhdanov, V.M., Aliyevskii, M.Ya.: Transfer and Relaxation Processes in Molecular Gases. Nauka, Moscow (1989) (in Russian)
3. Bertolotti, F.B.: The influence of rotational and vibrational energy relaxation on boundary-layer stability. J. Fluid Mech. **372**, 93–118 (1998)
4. Joseph, D.D., Carmi, S.: Stability of Poiseuille flow in pipes, annuli and channels. Quart. Appl. Math. **26**, 575–599 (1969)
5. Joseph, D.D.: Stability of Fluid Motion. Springer, Berlin (1976)
6. Malik, M., Dey, J., Alam, M.: Linear stability, transient energy growth, and the role of viscosity stratification in compressible plane Couette flow. Phys. Rev. E **77**, 036322(15) (2008)
7. Nagnibeda, E.A., Kustova, E.V.: Non-equilibrium Reacting Gas Flows. Kinetic Theory of Transport and Relaxation Processes. Springer, Berlin (2009)
8. Mack, L.M.: Boundary-layer stability theory. Report No. 900-277, Rev. A. Jet Propulsion Laboratory, Pasadena (1969)
9. Gaponov, S.A., Maslov, A.A.: Development of Perturbations in Compressible Flows. Nauka, Novosibirsk (1980) (in Russian)
10. Grigor'ev, Yu.N., Ershov, I.V.: Relaxation-induced suppression of vortex disturbances in a molecular gas. J. Appl. Mech. Tech. Phys. **44**, 471–481 (2003)
11. Grigor'ev, Yu.N.: On the energetic stability theory of compressible flows. Vychisl. Tekhnol. **11**, Special issue, 55–62 (2006) (in Russian)
12. Gol'dshtik, M.A., Shtern, V.N.: Hydrodynamic Stability and Turbulence. Nauka, Novosibirsk (1977) (in Russian)
13. Bronshtein, I.N., Semendyaev, K.A.: Reference Book on Mathematics. Nauka, Moscow (1986) (in Russian)
14. Korn, G.A., Korn, T.M.: Mathematical Handbook for Scientists and Engineers. McGraw-Hill, New York (1961)
15. Grigor'ev, Yu.N., Ershov, I.V.: On the effect of rotational relaxation on laminar-turbulent transition. In: Abstracts Conference Dedicated to the 40-year Anniversary of the Moscow State University (Moscow, November 22–26, 1999), pp. 65–66. Izd. Mosk. Gos. Univ., Moscow (1999) (in Russian)
16. Canuto, C., Hussaini, M.Y., Quarteroni, A., Zang, T.A.: Spectral Methods in Fluid Dynamics. Springer, Berlin (1988)
17. Trefethen, L.N.: Spectral Methods in Matlab. Society for Industrial and Applied Mathematics, Philadelphia (2000)
18. Moler, C.B., Stewart, G.W.: An algorithm for generalized matrix eigenvalue problems. SIAM J. Numer. Anal. **10**, 241–256 (1973)
19. Grigor'ev, Yu.N., Ershov, I.V.: Energy estimate of the critical Reynolds numbers in a compressible Couette flow. Effect of bulk viscosity. J. Appl. Mech. Tech. Phys. **51**, 669–675 (2010)
20. Grigor'ev, Yu.N., Ershov, I.V.: Effect of bulk viscosity on Kelvin-Helmholtz instability. J. Appl. Mech. Tech. Phys. **49**, 407–416 (2008)
21. Tamarkin, Ya.D.: Some General Problems of the Theory of Ordinary Differential Equations. Tipografiya M.P. Frolovoi, Petrograd (1917) (in Russian)
22. Naimark, M.N.: Linear Differential Operators. Nauka, Moscow (1969) (in Russian)
23. Grigor'ev, Yu.N., Ershov, I.V., Ershova, E.E.: Influence of vibrational relaxation on the pulsation activity in flows of an excited diatomic gas. J. Appl. Mech. Tech. Phys. **45**, 321–327 (2004)

Chapter 6
Evolution of a Large-Scale Vortex in Shear Flow of a Relaxing Molecular Gas

Abstract The chapter contains the results of the numerical study of a model problem for estimating the influence of thermal relaxation on the turbulized flow outside the limits of the laminar-turbulent transition. Nonlinear evolution of a large-scale vortex structure in a plane shear flow of a molecular gas is considered. Such structures are inevitable attributes of the final stage of the laminar-turbulent transition and turbulence generation in plane wakes, mixing layers, and submerged jets.

Modern scenarios of the laminar-turbulent transition show that origination, evolution, and decomposition of organized (coherent) vortex structures occur during the nonlinear stage of the process. The so-called λ-structures (horseshoes and hairpins) are observed in near-wall flows and tubes [1, 2]. In plane shear layers and jets two-dimensional vortices extended in the transverse direction are formed [3]. The same structures are observed in developed turbulent flows. Therefore, turbulence generation can be considered as laminar-turbulent transition process stochastically repeated in time and space. This allows us to assume that the role of relaxation of internal modes of molecules to the equilibrium state both at the nonlinear stage of the laminar-turbulent transition process and in developed turbulent flows can be estimated by modeling interaction of a solitary organized vortex structure with the stationary (mean) shear flow. Therefore, a two-dimensional model flow with the initial data being a superposition of the Rankine vortex and a linear shear flow is considered in this chapter. Despite its simplicity, this model can ensure an adequate reproduction of a typical feature of modern laminar-turbulent transition scenarios and turbulence generation in plane wakes and submerged jets, which is the evolution of large eddies on the background of the carrier shear flow. It may be also assumed that the heads of three-dimensional λ-structures evolve in the same manner.

Based on this model, the influence of thermal relaxation on suppression of vortex structures is considered for different levels of excitation of internal molecular modes.

Using the full Navier–Stokes equations for a compressible heat-conducting gas, a weakly nonequilibrium case is considered in Sect. 6.1, where the energy exchange between the translational and internal degrees of freedom of molecules is approximately described by the bulk viscosity coefficient in the stress tensor (1.19). An

Y.N. Grigoryev and I.V. Ershov, *Stability and Suppression of Turbulence in Relaxing Molecular Gas Flows*, Fluid Mechanics and Its Applications 117,
DOI 10.1007/978-3-319-55360-3_6

important factor in computations was the choice of model parameters, which allows approach to the model characteristics of those observed in real flows. The relative effect of additional suppression of disturbances associated with relaxation of weakly excited internal modes of gas molecules was preliminary estimated at a level of several percent. Therefore, the computational errors had to be within the third order of smallness. For this purpose, the difference scheme used in numerical calculations was carefully tested.

The results of calculations of the model flow on the background of relaxation of a vibrationally excited gas are reported in Sect. 6.2. The evolution of the vortex structure is described by a system of equations of two-temperature aero-gas dynamics (1.31). The parameters of vibrational excitation are chosen within realistic ranges for real diatomic gases, where dissociation can be neglected.

For estimating the influence of relaxation of vibrationally excited internal modes of gas molecules on the fluctuating characteristics of the model flow, the time evolution of the total kinetic energy and the absolute value of the Reynolds stresses is considered in both cases.

6.1 Navier–Stokes Model Flow. Effect of Bulk Viscosity

6.1.1 Parametrization of a Model Flow

Here we consider the influence of bulk viscosity on the laminar-turbulent transition and turbulence generation. The object for study is a two-dimensional vortex structure embedded in a plane shear flow. Such structures emerge in plane mixing layers, jets, and wakes of bodies in the flow. The Mach number in the main flow is assumed to be $M < 1$. In this case the compressibility influence on kinematic and dynamic characteristics of the structure can be neglected, and modeling parameters can be chosen using validated data on mixing layers [3, 4].

The relative intensity of a vortex structure is given by the parameter

$$\beta = u'/\Delta u \approx v'/\Delta u$$

(u' and v' are the maximum velocity fluctuations and Δu is the absolute difference in velocities on the mixing layer boundaries). According to [3] (see also [4]),

$$\beta \simeq 0.2\text{–}0.5.$$

Intermittency in the streamwise direction is given by the relation of the vortex size R_0 to the half-distance between the structures $l/2$. On the average, for mixing layers and jets, the intermittency coefficient is

$$\chi = \frac{l}{2R_0} = 2\text{–}6.$$

The characteristic relation between the layer thickness δ and the structure size is estimated as [3, 4]

$$\frac{\delta}{R_0} \simeq 2\text{–}4.$$

It follows from experimental data and calculations (see [3] and references to it) that the dynamics of all disturbances, including large-scale structures, is universal in mixing layers for the Reynolds numbers

$$\mathrm{Re} = \frac{\rho \Delta u \delta}{\eta} \geq 10^2.$$

For this reason calculations can be performed for moderate Reynolds numbers, thus, avoiding certain computational difficulties.

In choosing the velocity profile of the carrier flow the Tollmien solution [4], can be applied to the mixing layer on the jet boundary. In self-similar variables, the streamwise velocity profile in the layer has the form

$$\overline{u}_T = u(\varphi)/(2U_0) = F'(\varphi)$$

$$= 0.0176\ \exp(-\varphi) + 0.6623\ \cos(\sqrt{3}\varphi/2) + 0.2280\ \exp(\varphi)\sin(\sqrt{3}\varphi/2), \qquad (6.1)$$

where $\varphi = x_2/(ax_1)$ and a is an empirical constant characterizing the layer structure. The function $F'(\varphi)$ is tabulated in [4]. For $\varphi_0 = 0.98$, profile (6.1) is conjugated with the jet core, where $F'(\varphi_0) = 1$ and $u(\varphi_0) = 2U_0$. The value of $\varphi_n = -2.04$ corresponds to the external boundary of the jet, where $F'(\varphi_n) = 0$. The linear velocity profile between these values is given by the formula

$$\overline{u}_l = \frac{u}{2U_0} = \frac{\varphi - \varphi_n}{\varphi_0 - \varphi_n}. \qquad (6.2)$$

A comparison of profiles (6.1) and (6.2) shows that the root-mean-square deviation

$$\sigma = \sqrt{\frac{1}{n}\sum_{i=1}^{n}(\bar{u}_T(\varphi_i) - \bar{u}_l(\varphi_i))^2}$$

does not exceed $\sigma = 0.075$. This proves the feasibility of modeling the carrier stream by a linear shear flow.

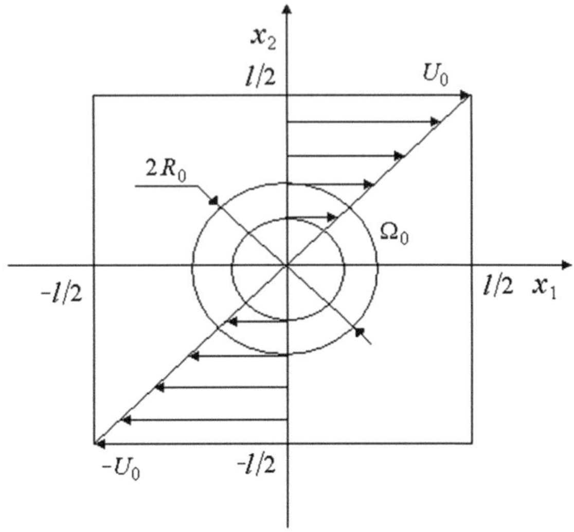

Fig. 6.1 Flow pattern at the initial time

If the structure center is considered to be located on the mid-streamline and moves with a velocity U_0, we can pass to the system of coordinates where it is at rest. The Rankine vortex of radius R_0 with a constant vorticity density Ω_0 is used as an initial state of the vortex structure. Figure 6.1 shows the thus symmetrized initial flow in a model cell.

The calculations are performed for the following values of the parameters:

$$\alpha_1 = \frac{\eta_b}{\eta} = 0\text{–}2, \quad \beta = \frac{\Omega_0 R_0}{2U_0} = 0.2\text{–}0.5, \quad \chi = \frac{l}{2R_0} = 2\text{–}6,$$

$$\mathrm{M} = \frac{U_0}{\sqrt{\gamma R T_0}} = 0.2\text{–}0.8, \quad \mathrm{Re} = \frac{2U_0 R_0 \rho_0}{\eta} = 40\text{–}100, \quad \mathrm{Pr} = \frac{\eta c_p}{\lambda} = 0.74.$$

6.1.2 Basic Equations and Initial-Boundary Conditions

The structure evolution in the model cell is described by a system of full Navier–Stokes equations for a compressible viscous heat-conducting gas. Calculations make it possible to analyze energy and momentum exchange between the disturbance and the main flow, which occurs both at the loss of stability and in the course of turbulence generation. The reference values chosen for normalization are the initial diameter of the structure $2R_0$, the absolute value of velocity U_0, the density ρ_0, the temperature T_0 on the upper and lower boundaries of the model cell, the time $\tau_0 = 2R_0/U_0$, and the pressure $p_0 = \rho_0 U_0^2$.

In dimensionless variables the system of equations is written as

$$\frac{\partial \rho}{\partial t} + \frac{\partial \rho u_i}{\partial x_i} = 0,$$

$$\rho\left(\frac{\partial u_i}{\partial t} + u_j \frac{\partial u_i}{\partial x_j}\right) = -\frac{\partial p}{\partial x_i} + \frac{1}{\mathrm{Re}}\frac{\partial^2 u_i}{\partial x_j^2} + \frac{1}{\mathrm{Re}}\left(\alpha_1 + \frac{1}{3}\right)\frac{\partial^2 u_j}{\partial x_i\,\partial x_j},$$

$$\rho\left(\frac{\partial T}{\partial t} + u_i \frac{\partial T}{\partial x_i}\right) = (\gamma - 1)\mathrm{M}^2 \frac{dp}{dt} + \frac{1}{\mathrm{Re}\,\mathrm{Pr}}\frac{\partial^2 T}{\partial x_i^2}$$

$$+\frac{(\gamma - 1)\mathrm{M}^2}{2\mathrm{Re}}\left(\frac{\partial u_i}{\partial x_j} + \frac{\partial u_j}{\partial x_i}\right)^2 + (\gamma - 1)\left(\alpha_1 - \frac{2}{3}\right)\frac{\mathrm{M}^2}{\mathrm{Re}}\left(\frac{\partial u_i}{\partial x_i}\right)^2,$$

$$\gamma \mathrm{M}^2 p = \rho T, \quad \frac{d}{dt} = \frac{\partial}{\partial t} + u_i \frac{\partial}{\partial x_i}, \quad i,\ j = 1,\ 2. \tag{6.3}$$

Here all notations of variables are the same as used in (5.1).

On the cell boundaries the following conditions are satisfied at all times: for $x_1 = \pm\chi/2$ and $x_2 \in [-\chi/2;\ \chi/2]$,

$$u_1(t,\ \chi/2,\ x_2) = u_1(t,\ -\chi/2,\ x_2), \quad u_2(t,\ \chi/2,\ x_2) = -u_2(t,\ -\chi/2,\ x_2),$$

$$\rho(t,\ \chi/2,\ x_2) = \rho(t,\ -\chi/2,\ x_2), \quad p(t,\ \chi/2,\ x_2) = p(t,\ -\chi/2,\ x_2); \tag{6.4}$$

for $x_2 = \pm\chi/2$ and $x_1 \in [-\chi/2;\ \chi/2]$,

$$u_1(t,\ x_1,\ \chi/2) = -u_1(t,\ x_1,\ -\chi/2), \quad u_2(t,\ x_1,\ \chi/2) = u_2(t,\ x_1,\ -\chi/2),$$

$$\rho(t,\ x_1,\ \chi/2) = \rho(t,\ x_1,\ -\chi/2), \quad p(t,\ x_1,\ \chi/2) = p(t,\ x_1,\ -\chi/2). \tag{6.5}$$

The carrier flow in the calculation domain is specified as an exact stationary solution of system (6.3) with boundary values depending on x_1 and x_2 similarly to (6.4) and (6.5). In dimensionless variables the profiles of velocity U_2, temperature T, and density are written as

$$U_2(x_2) = \frac{2x_2}{\chi}, \quad T(x_2) = 1 + \frac{(\gamma - 1)\mathrm{M}^2\,\mathrm{Pr}}{2}\left(1 - \frac{4x_2^2}{\chi^2}\right),$$

$$\Theta(x_2) = T^{-1}(x_2). \tag{6.6}$$

At the same time, as follows from (6.3) and from the equation of state, the pressure in such a flow is constant over the space:

$$P = \frac{1}{\gamma\,\mathrm{M}^2}.$$

The initial conditions for the velocity field are given in the form

$$u_1(0, x_1, x_2) = \begin{cases} 2x_2/\chi + \beta x_2/(2r^2), & r > 1/2, \\ 2x_2/\chi + 2\beta x_2, & r \leq 1/2; \end{cases}$$

$$u_2(0, x_1, x_2) = \begin{cases} -\beta x_1/(2r^2), & r > 1/2, \\ -2\beta x_1, & r \leq 1/2, \end{cases} \tag{6.7}$$

where $r = \sqrt{x_1^2 + x_2^2}$. The initial distributions of thermodynamic quantities correspond to the conditions of an undisturbed carrier flow (6.6).

In calculating the nonlinear interaction of the disturbance with the main flow integral equations of balance between bilinear fluctuating characteristics and the kinetic energy of the disturbance are used. To derive these equations within the scope of this model problem instantaneous values of hydrodynamic quantities are presented as

$$u_i = U_i + u_i', \quad p = P + p', \quad \rho = \Theta + \rho'. \tag{6.8}$$

In (6.8) a stationary solution U_i, P, and Θ of the system of Navier–Stokes equations (carrier flow) and the imposed disturbance satisfying conditions (6.4) and (6.5) are singled out. The carrier flow is assumed to be one-dimensional and depend only on the transverse coordinate x_2: $U_1 = U_1(x_2)$. It is supposed that the velocity disturbance in the vicinity of the calculation cell satisfies the conditions

$$u_1'(t, x_1, x_2) = u_1'(t, -x_1, x_2), \quad u_2'(t, x_1, x_2) = -u_2'(t, -x_1, x_2),$$

$$u_1'(t, x_1, x_2) = -u_1'(t, x_1, -x_2), \quad u_2'(t, x_1, x_2) = u_2'(t, x_1, -x_2). \tag{6.9}$$

By substituting (6.8) into system (6.3) and by eliminating terms related to the stationary solution, we obtain the following equations for disturbances of density and velocities with unrestricted amplitudes:

$$\frac{\partial \rho'}{\partial t} + u_i \frac{\partial \rho}{\partial x_i} + \rho \frac{\partial u_i'}{\partial x_i} = 0, \tag{6.10}$$

$$\rho \left(\frac{\partial u_i'}{\partial t} + u_j' \frac{\partial u_i'}{\partial x_j} + U_j \frac{\partial u_i'}{\partial x_j} + u_j' \frac{\partial U_i}{\partial x_j} \right) + \rho' U_j \frac{\partial U_i}{\partial x_j}$$

$$= -\frac{\partial p'}{\partial x_i} + \frac{1}{\mathrm{Re}} \frac{\partial^2 u_i'}{\partial x_j^2} + \frac{1}{\mathrm{Re}} \left(\alpha_1 + \frac{1}{3} \right) \frac{\partial^2 u_j'}{\partial x_i \, \partial x_j}, \quad i, \, j = 1, \, 2. \tag{6.11}$$

By multiplying Eq. (6.11) by the fluctuating component of velocity u_k' and performing symmetrization over indices, we obtain the equation for bilinear fluctuations

$$\frac{\partial}{\partial t}\left(\rho u_i' u_k'\right) + \frac{\partial}{\partial x_j}\left(\rho u_j' u_i' u_k'\right) + \frac{\partial}{\partial x_j}\left(\rho u_i' u_k' U_j\right)$$

$$+\left(\rho u_j' + \rho' U_j\right)\left(u_k' \frac{\partial U_i}{\partial x_j} + u_i' \frac{\partial U_k}{\partial x_j}\right) = -\left(u_k' \frac{\partial p'}{\partial x_i} + u_i' \frac{\partial p'}{\partial x_k}\right)$$

$$+\frac{1}{\mathrm{Re}}\left(u_k' \frac{\partial^2 u_i'}{\partial x_j^2} + u_i' \frac{\partial^2 u_k'}{\partial x_j^2}\right) + \frac{1}{\mathrm{Re}}\left(\alpha_1 + \frac{1}{3}\right)\left(u_k' \frac{\partial}{\partial x_i} + u_i' \frac{\partial}{\partial x_k}\right)\frac{\partial u_j'}{\partial x_j}. \tag{6.12}$$

On the left side of Eq. (6.12) terms in the divergent form are found using the continuity equation (6.10). By integrating Eq. (6.12) over the calculation domain with allowance for conditions (6.4), (6.5), and (6.9) one obtains the integral equation for bilinear fluctuating characteristics

$$\frac{d}{dt}\int_{\Omega}\left(\rho u_i' u_k'\right) d\Omega = -\int_{\Omega} \rho u_j' \left(u_i' \frac{\partial U_k}{\partial x_j} + u_k' \frac{\partial U_i}{\partial x_j}\right) d\Omega$$

$$+\int_{\Omega} p' \left(\frac{\partial u_k'}{\partial x_i} + \frac{\partial u_i'}{\partial x_k}\right) d\Omega - \frac{2}{\mathrm{Re}}\int_{\Omega}\left(\frac{\partial u_i'}{\partial x_j}\frac{\partial u_k'}{\partial x_j}\right) d\Omega$$

$$-\frac{1}{\mathrm{Re}}\left(\alpha_1 + \frac{1}{3}\right)\int_{\Omega}\left(\frac{\partial u_k'}{\partial x_i} + \frac{\partial u_i'}{\partial x_k}\right)\frac{\partial u_j'}{\partial x_j} d\Omega, \quad i,\ j,\ k = 1,\ 2. \tag{6.13}$$

By convolving Eq. (6.13) over indices for $i = k$ one derives an integral equation of balance for the kinetic energy of disturbances considered in previous Chap. 5.

$$\frac{dE}{dt} \equiv \frac{d}{dt}\int_{\Omega} \frac{\rho u_i'^2}{2} d\Omega = J_1 + J_2 - \frac{1}{\mathrm{Re}}\left(J_3 + \alpha_1 J_4\right). \tag{6.14}$$

If averaging over the cell area $|\Omega|$

$$\langle G \rangle = \frac{1}{|\Omega|}\int_{\Omega} G\, d\Omega$$

is introduced, then Eqs. (6.13) and (6.14) can be treated with accuracy up to a constant factor as equations for averaged characteristics.

As we marked above the integrals J_3 and J_4 are nonnegative. It follows from Eq. (6.14) that the bulk and dynamic viscosities promote disturbance suppression in the case of nonlinear interaction of the structure with the main flow.

6.1.3 Numerical Calculations of a Model Flow

Difference scheme. In numerical calculations system (6.3) is approximated by a weighting finite-difference scheme with splitting in terms of spatial coordinates and physical processes [5]. In the operator form the scheme is written as

$$(\mathbf{x}^{n+1} - \mathbf{x}^n)/\tau + L_h[\delta \mathbf{x}^{n+1} + (1-\delta)\mathbf{x}^n] = \mathbf{F}_h^n. \tag{6.15}$$

Here

$$\mathbf{x}^n = (\rho_{ij}^n,\ u_{1,ij}^n,\ u_{2,ij}^n,\ T_{ij}^n)$$

is the grid vector function of the solution in the nth time layer, h is the grid spacing, τ is the time step, and δ is the weighting parameter. The operator L_h includes symmetric second-order approximations of the first and second spatial derivatives along each spatial coordinate. The operator $\mathbf{F}_h^n$ is treated as a vector of right sides and is composed of second-order approximations symmetric about each coordinate of mixed derivatives from equations of momenta and the terms of the dissipative function from the energy equation. On a regular grid with a step h along both coordinates scheme (6.15) approximates system (6.3) with an order $O(\tau + h^2)$ and is absolutely stable for the weighting parameter $\delta > 1/2$ [5].

Test calculations. The effect of additional disturbance suppression due to bulk viscosity was preliminarily estimated as several percent. Therefore, the calculation errors have to be lower than the third order of smallness. To reach such a result the difference scheme was carefully tested. In the test calculations the next values of flow parameters were chosen M $= 0.5$–0.8, Re $= 100$, Pr $= 0.74$, $\beta = 0.2$, $\chi = 3$, $\gamma = 1, 4$. A sequence of embedded grids with spacings h =0.1, 0.05, 0.025, and 0.0125 was used. The time step was $\tau = 0.01$. Disturbances considered were defined as difference between actual values of hydrodynamic characteristics and corresponding ones of stationary flow (6.8). Scheme (6.15) had to retain a stationary carrier flow (6.6) in the absence of an imposed disturbance. In test calculations at the times (600–800)τ, deviations from the stationary profiles of hydrodynamic quantities in fact did not exceed the rounding error, remaining at the level $\varepsilon_s = 10^{-6}$.

Additionally, in the absence of the carrier flow, the difference problem (6.15) was solved with the initial-boundary conditions (6.4), (6.5), and (6.7). By virtue of the problem symmetry the velocity-field divergence is zero. On the assumption of an isochoric process with $\rho =$ const the continuity equation is satisfied identically. In this case the evolution of the initial disturbance in the form of the Rankine vortex is described by the well-known analytical solution for a viscous incompressible fluid [6]. According to the introduced notation, the expression for the vorticity field has the form

$$\omega(r,\, t) = \frac{\beta\chi \mathrm{Re}}{t} \exp\left(-\frac{r^2 \mathrm{Re}}{4t}\right) \int_0^{1/2} \exp\left(-\frac{\xi^2 \mathrm{Re}}{4t}\right) I_0\left(\frac{r\mathrm{Re}}{2t}\,\xi\right) \xi \, d\xi, \tag{6.16}$$

where $I_0(\zeta)$ is the zero-order Bessel function of the second kind of the imaginary argument and

$$r = \sqrt{x_1^2 + x_2^2}$$

is the distance from the vortex center to an arbitrary point in the calculation cell.

In the numerical solution the grid vorticity function is calculated on the basis of the symmetrical finite-difference approximation

$$\omega_{ij}^n = \frac{\chi}{2}\left(\frac{u_{1,i,j-1}^n - u_{1,i,j+1}^n}{2h} - \frac{u_{2,i-1,j}^n - u_{2,i+1,j}^n}{2h}\right). \tag{6.17}$$

The maximum difference between the calculated values (6.17) and the analytical solution (6.16) in all grid nodes did not exceed the value $\varepsilon_d = 5 \times 10^{-3}$ in the time interval up to 600τ for the range of parameters examined.

For testing the scheme accuracy for $\rho \neq \text{const}$, the coincidence between the time evolution of vorticity at the center of the Rankine vortex in compressible and incompressible viscous fluids was used. The respective analytical solution for the vorticity is given by the formula [6]

$$\omega(0,t) = 2\beta\chi\left(1 - \exp\left(-\frac{\text{Re}}{16t}\right)\right).$$

As the calculation shows, in this case the maximum difference of the numerical solution at the center of the model cell from this analytical expression does not exceed the value $\varepsilon_d = 5 \times 10^{-3}$ in the time interval of up to 600τ.

The test results confirm that the numerical model developed ensures sufficient accuracy of the problem solution.

Calculation results of a model flow. The energy exchange between the carrier flow and disturbances of various space and time scales is one of the basic processes both in the laminar-turbulent transition and in developed turbulence. To estimate the influence of bulk viscosity on fluctuating characteristics of the model flow, we investigated the time evolution of absolute values of the Reynolds stresses

$$\sigma_{12}(t) = \int_{-\chi/2}^{\chi/2}\int_{-\chi/2}^{\chi/2} \left|\rho u_1' u_2'\right| dx_1\, dx_2 \tag{6.18}$$

and the kinetic energy of disturbances

$$E(t) = \frac{1}{2}\int_{-\chi/2}^{\chi/2}\int_{-\chi/2}^{\chi/2} \rho\left(u_1'^2 + u_2'^2\right) dx_1\, dx_2. \tag{6.19}$$

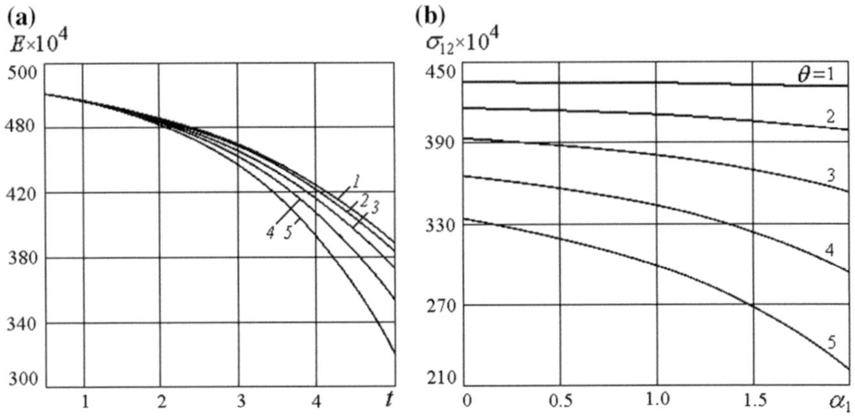

Fig. 6.2 Effect of bulk viscosity disturbance evolution (Re $=$ 100, Pr $= 0.74$, M $= 0.5$, $\beta = 0.2$, $\chi = 3$, $\gamma = 1.4$). **a** shows kinetic energy versus time for $\alpha_1 = 0$ (*1*), 0.5 (2), 1 (*3*), 1.5 (*4*), and 2 (*5*). **b** shows absolute values of the Reynolds stresses σ_{12} versus the parameter α_1 at different times θ

The integrands in (6.18) and (6.19) were calculated from instantaneous flow characteristics obtained from the numerical solution of system (6.3) using the difference scheme (6.15). The integrals were calculated by the rectangle formula on a regular grid with a step $h = 0.1$.

An example of calculating the dependencies $E(t)$ for $\alpha_1 = 0\text{–}2$ is shown in Fig. 6.2a. The dependencies $\sigma_{12}(t)$ have a similar character. Figure 6.2b shows the curves $\sigma_{12}(\alpha_1)$ at various times θ. On averaging over the same time intervals in the form,

$$\langle F(\alpha_1)\rangle = \frac{1}{\theta}\int_0^\theta F(t,\alpha_1)\,dt,$$

the dependencies $\langle E(\alpha_1)\rangle$ and $\langle\sigma_{12}(\alpha_1)\rangle$ become less steep than those in Fig. 6.2b, but their character is retained. The order of magnitude of the time interval up to $\theta = 5$ corresponds to the average lifetime of a large structure. At this moment in the calculations the disturbance level on the boundary reached the value of the numerical calculation error, after which the computation was stopped.

As follows from the presented plots, as the coefficient of bulk viscosity grows, the kinetic energy of disturbances and the Reynolds stresses decay more intensively. The maximum stratification of the curves $E(t)$ and the most drastic change in $\sigma_{12}(\alpha_1)$ are observed at $\theta = 4\text{–}5$. In order to estimate the effect of bulk viscosity on the average fluctuating characteristics, the relative changes

$$\Delta_F = \left|\frac{\langle F(\alpha_1)\rangle - \langle F(0)\rangle}{\langle F(0)\rangle}\right|, \quad \alpha_1 = 0\text{–}2 \tag{6.20}$$

for the time interval $0 \leq \theta \leq 5$ were calculated. For the calculation conditions corresponding to Fig. 6.2, both for the kinetic energy $\langle E(\alpha_1)\rangle$ and for the absolute value of the Reynolds stresses $\langle\sigma_{12}(\alpha_1)\rangle$, the characteristic Δ_F changes in the range from 0.012 to 0.096 for $\alpha_1 = 0.5$–2. As the Mach number increases, both limits proportionately grow, and the upper limit reaches the value $\Delta_F \simeq 0.2$ for $\mathrm{M} = 2$. However, for this value of M, the above-chosen parametrization of the model flow, and the model of bulk viscosity itself, can be unsuitable.

In the mixing layer the contribution of organized vortices to the total Reynolds stresses and kinetic energy of fluctuations is known to be approximately 40% [3]. It is obvious, at the same time, that molecular dissipation suppresses small-scale disturbances much more intensively than large-scale disturbances considered here. Proceeding from this fact, the calculated changes in the value of Δ_F $(\Delta_F \leq 0, 1)$ may be extended to the entire spectrum of disturbances.

On the basis of the discrete approximation of Eq. (6.14) the time evolution of the kinetic energy $dE(t)/dt$ is calculated. The respective plots are shown in Fig. 6.3. To test the calculation accuracy the finite-difference approximation

$$\frac{dE}{dt} \simeq \frac{E(t+\tau) - E(t-\tau)}{2\tau} \tag{6.21}$$

is calculated parallel to the calculations by Eq. (6.14). Scattering of the data obtained by the two methods does not exceed 1%. It is obvious that, with increasing parameter α_1, the absolute value of the kinetic energy dissipation rate increases. The dissipation rate also grows with time. The time-averaged dependencies $\langle dE/dt\rangle$ (Fig. 6.3a) have lower gradients than the curves in Fig. 6.3b; at the same time the character of the

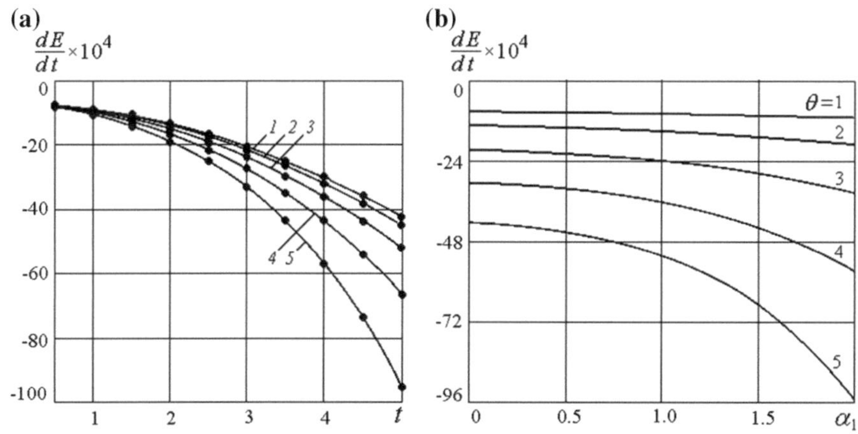

Fig. 6.3 Generation of kinetic energy of disturbances. **a** shows versus time; the curves are constructed on the basis of the calculation results of Eq. (6.14); the points refer to the calculation results of Eqs. (6.19) and (6.21). **b** shows versus bulk viscosity at different times θ; the regime parameters and notation employed are the same as in Fig. 6.2

dependencies remains unchanged. As the Mach number grows, the influence of bulk viscosity on the dissipation rate becomes stronger. We can state that the relative differences for the averaged values of $\langle dE/dt \rangle$ calculated from relation (6.14) are within the same limits as for the kinetic energy of disturbances.

By such a way using a simple model, the influence of bulk viscosity on the nonlinear interaction of a finite-amplitude vortex disturbance with a carrier shear flow was studied. The range of parameters of the model flow correspond to the real values for diatomic gases. Numerical modeling results seem to evidence a stabilizing effect of bulk viscosity on the dynamics of disturbances. It was shown that, in a weakly compressible flow with $\mathrm{M} < 1$, as bulk viscosity increases in the range $0 \leq \eta_b \leq 2\eta$, the relative decrease in the absolute value of the Reynolds stresses and kinetic energy of disturbances increase and approach 10% for $\eta_b = 2\eta$. The results obtained are not final and require further investigation based on improved and more perfect models. It should be noted, that such a suppression of fluctuating intensity can be achieved by means of well-known mechanical methods for drag reduction, for instance, by ribleting surfaces in the flow [7]. Therefore, the results point to a hypothetical possibility of drag control in compressible flows by regulating bulk viscosity of the gas with use of some physical-chemical processes.

An analysis of the calculation results shows that generation of kinetic energy of fluctuations in this model problem is always negative. The absence of a mechanism for positive generation of disturbance energy, for instance, such as stretching of quasi-streamwise vortices connecting large structures in the mixing layer [3], is a drawback of a simplified two-dimensional model preventing its application to immediate estimation of the effect of bulk viscosity η_b on the critical transition Reynolds number.

6.2 Effect of a Vibrational Relaxation on Damping Vortex Structure

In this subsection suppression of large vortex structure by relaxation of vibrational mode is considered. The same model flow as in Sect. 6.1 is calculated with use of the complete system of equations of two-temperature gas dynamics for a viscous heat-conducting gas.

6.2.1 Basic Equations and Initial-Boundary Conditions

The equations were normalized by the same characteristic values as used in (6.3). In the dimensionless variables the system of equations is written as follows

$$\frac{\partial \rho}{\partial t} + \frac{\partial \rho u_i}{\partial x_i} = 0,$$

$$\rho\left(\frac{\partial u_i}{\partial t}+u_j\frac{\partial u_i}{\partial x_j}\right)=-\frac{\partial p}{\partial x_i}+\frac{1}{\mathrm{Re}}\frac{\partial^2 u_i}{\partial x_j^2}+\frac{1}{\mathrm{Re}}\left(\alpha_1+\frac{1}{3}\right)\frac{\partial^2 u_j}{\partial x_i\,\partial x_j},$$

$$\rho\left(\frac{\partial T}{\partial t}+u_j\frac{\partial T}{\partial x_j}\right)+\gamma(\gamma-1)\mathrm{M}^2 p\frac{\partial u_i}{\partial x_i}=\frac{\gamma}{\mathrm{Re\,Pr}}\frac{\partial^2 T}{\partial x_i^2}+\frac{\gamma_v\,\rho\,(T_v-T)}{\tau_{vt}}$$

$$+\frac{\gamma(\gamma-1)\mathrm{M}^2}{2\mathrm{Re}}\left[\left(\frac{\partial u_i}{\partial x_j}+\frac{\partial u_j}{\partial x_i}\right)^2+2\left(\alpha_1-\frac{2}{3}\right)\left(\frac{\partial u_i}{\partial x_i}\right)^2\right],$$

$$\gamma_v\,\rho\left(\frac{\partial T_v}{\partial t}+u_j\frac{\partial T_v}{\partial x_j}\right)=\frac{\gamma\,\alpha_2}{\mathrm{Re\,Pr}}\frac{\partial^2 T_v}{\partial x_i^2}-\frac{\gamma_v\,\rho\,(T_v-T)}{\tau_{vt}},$$

$$\gamma \mathrm{M}^2 p=\rho T,\quad i,j=1,2. \tag{6.22}$$

In system (6.22) the equation for the vibrational temperature T_v is a relaxation Landau-Teller equation written in Euler form [8, 9]. The first term on the right side of this equation describes the local flux of vibrational energy due to molecular transfer, and the second term describes the vibrational energy relaxation to equilibrium.

On the mesh boundaries at all times the conditions (6.4) and (6.5) were specified. In addition for $x_1=\pm\chi/2$ and $x_2\in[-\chi/2;\ \chi/2]$ we used the next conditions for vibrational temperature:

$$T_v(\chi/2,x_2,t)=T_v(-\chi/2,x_2,t); \tag{6.23}$$

$$T_v(x_1,\ \chi/2,\ t)=T_v(x_1,\ -\chi/2,\ t). \tag{6.24}$$

In the calculation domain the carrier flow region is specified as the exact steady-state solution of system (6.22) for the case of thermal equilibrium over the degrees of freedom with boundary conditions (6.23) and (6.24). In the dimensionless variables the flow characteristics have the form

$$U(x_2)=\frac{2x_2}{\chi},\quad V=0,\quad T_0(x_2)=T_{v,0}(x_2)=1+\frac{(\gamma-1)\mathrm{M}^2\mathrm{Pr}}{2}\left(1-\frac{4x_2^2}{\chi^2}\right),$$

$$\rho_0(x_2)=\frac{1}{T_0(x_2)},\quad P_0(x_1,\ x_2)=\frac{1}{\gamma\ \mathrm{M}^2}. \tag{6.25}$$

The initial conditions for the velocity field and thermodynamic quantities were specified as

$$u(0,\ x_1,\ x_2)=\begin{cases}U(x_2)+\beta x_2/(2r^2), & r>1/2,\\ U(x_2)+2\beta x_2, & r\le 1/2,\end{cases}$$

$$v(0,\ x_1,\ x_2)=\begin{cases}-\beta x_1/(2r^2), & r>1/2,\\ -2\beta x_1, & r\le 1/2,\end{cases}$$

$$T(0,\, x_1,\, x_2) = T_0(x_2), \quad T_v(0,\, x_1,\, x_2) = (1+\xi)T_0(x_2),$$

$$\rho(0,\, x_1,\, x_2) = \rho_0(x_2), \quad r = \sqrt{x_1^2 + x_2^2}.$$

The calculations were performed for the following parameter values: Mach number $\mathrm{M} = U_0/\sqrt{\gamma R T_0} = 0.5$, Reynolds number $\mathrm{Re} = 2U_0 R_0 \rho_0/\eta = 100$, Prandtl number $\mathrm{Pr} = \eta c_p/\lambda = 0.74$, intermittency parameter $\chi = l/(2R_0) = 3$, relative intensity of vortex perturbation $\beta = \Omega_0 R_0/(2U_0) = 0.2$, $\gamma = 1.4$, $\alpha_1 = 0$–2, $\alpha_2 = 0.2424$, $\xi = 0$–5, $\tau_{vt} = 0$–5.

6.2.2 *Numerical Scheme and Results of Calculations*

In the numerical calculations system (6.22) was approximated by a weight finite-difference scheme with splitting in physical processes and space coordinates similar to that used in Sect. 6.1. The scheme is written in operator form as

$$\frac{\mathbf{x}^{n+1} - \mathbf{x}^n}{\Delta t} + \mathbf{L}_h[\delta \mathbf{x}^{n+1} + (1-\delta)\mathbf{x}^n] = \mathbf{G}_h^n, \tag{6.26}$$

where

$$\mathbf{x}^n = (\rho_{ij}^n,\, u_{ij}^n,\, v_{ij}^n,\, T_{ij}^n,\, T_{v,ij}^n)$$

is a grid vector function of the solution in the nth time layer, h and Δt are the spatial grid step and the step in time, respectively, and δ is the weight parameter. The form of the operator $\mathbf{L}_h$ is described in Sect. 6.1 and the right side vector $\mathbf{G}_h^n$ consists of second-order approximations symmetric over each coordinate with mixed derivatives from the momentum equations, the terms of the dissipative function from the energy equation, and the source terms from the relaxation equation and energy equation (see also [10]). In the quasi-equilibrium case where system (6.22) becomes the complete system of Navier–Stokes equations, scheme (6.26) coincides with the scheme proposed in [5] (see also Sect. 6.1). An analysis shows that the appearance of an additional equation and a source term in the energy equation does not change its computational properties. Thus, scheme (6.26) on a regular grid with a step h on both coordinates approximates system (6.22) with order $O(\Delta t + h^2)$ and is absolutely stable for a weight parameter $\delta > 1/2$.

To estimate the contribution of the relaxation process, the model problem in question was also solved in an approximation of relaxation gas dynamics, in which the momentum and energy equations and the relaxation equation of system (6.22) do not contain dissipative terms ($\eta = \eta_b = 0$ and $\lambda_t = \lambda_r = \lambda_v = 0$). In this case the system of equations was approximated by a weight finite-difference scheme with splitting in physical processes and space coordinates. It was constructed on the basis of a numerical scheme for the equations of gas dynamics from [5]. In abstract operator form the scheme is also written in the form (6.26). However, in this scheme the operator $\mathbf{L}_h$ is composed of symmetric second-order approximations of the first derivatives with respect to each space coordinate, and the right-side vector $\mathbf{G}_h^n$ consists of the relaxation terms from the energy equation and the relaxation equation. The approximation order and the stability of the thus obtained scheme are similar to the characteristics of scheme (6.26).

The evolution of the perturbation was traced up to its entry to the mesh boundary, which required up to 600 time steps, after which the calculation was terminated.

To improve the numerical schemes, we performed test calculations similar to those carried out in Sect. 6.1. As was shown, the maximum calculation errors do not exceed 5×10^{-3}.

In the calculations we examined the energy and momentum transfer between the imposed perturbation and the carrier flow. The pulsation flow characteristics Φ' were defined by

$$\Phi' = \Phi - \Phi_0,$$

where Φ are instantaneous values of the flow characteristics obtained by numerical solution of the model problem and Φ_0 are the characteristics of the equilibrium steady-state carrier flow (6.25). To estimate the effect of relaxation of excited vibrational degrees of freedom and bulk viscosity on the pulsation characteristics of the model flow, we examined the time evolution of the Reynolds stress modulus

$$\sigma_{12}(t) = \int_{-\chi/2}^{\chi/2} \int_{-\chi/2}^{\chi/2} \left|\rho u' v'\right| dx_1\, dx_2$$

and the kinetic energy of perturbations

$$E(t) = \frac{1}{2} \int_{-\chi/2}^{\chi/2} \int_{-\chi/2}^{\chi/2} \rho \mathbf{u}'^2\, dx\, dy.$$

The corresponding integrals were evaluated using the rectangle formulas on a regular grid with a step $h = 0.1$.

In Fig. 6.4a the plots of $E(t)$ demonstrate the effect of the purely relaxation process versus the degree of excitation of the vibrational mode ξ for the case where, in Eq. (6.22), the dissipative coefficients are equal to zero ($\eta = \eta_b = 0$ and $\lambda_t = \lambda_r = \lambda_v = 0$) and the vibrational relaxation time $\tau_{vt} = 3$ is comparable to the characteristic time of flow evolution. The course of the curves shows that as the value of ξ increases, the dependencies $E(t)$ decrease more rapidly. The effect of the variation in the vibrational relaxation time τ_{vt} on the damping of the perturbation energy $E(t)$ is illustrated in Fig. 6.4b. The parameter τ_{vt} varies within one order of magnitude. The values of the parameters α_1 and ξ, which also influence dissipation, are moderate. It is obvious that with decrease in the relaxation time, the pulsations are damped more rapidly, although not in a direct proportion to the variation of τ_{vt}. The dependencies $\sigma_{12}(t)$ for various ξ and τ_{vt} behave the same as the curves presented in the figure. As the degree of excitation of the vibrational energy levels ξ increases, the Reynolds stresses damp more rapidly, and a decrease in the vibrational relaxation time τ_{vt} leads to an increase in the damping rate $\sigma_{12}(t)$.

The effect of the vibrational relaxation process on the mean pulsation characteristics was estimated quantitatively by calculating their relative changes:

$$\Delta_F^{\xi} = \left|\frac{\langle F(\xi, \tau_{vt})\rangle - \langle F(0, \tau_{vt})\rangle}{\langle F(0, \tau_{vt})\rangle}\right|, \quad \tau_{vt} = \text{const},$$

$$\Delta_F^{\tau_{vt}} = \left|\frac{\langle F(\xi, \tau_{vt})\rangle - \langle F(\xi, 0)\rangle}{\langle F(\xi, 0)\rangle}\right|, \quad \xi = \text{const}.$$

The averaging in time was defined by

$$\langle F\rangle = \frac{1}{\theta}\int_0^{\theta} F(t)\,dt.$$

For the regime shown in Fig. 6.4 in an approximation of relaxation gas dynamics with $\eta = \eta_b = 0, \lambda_t = \lambda_r = \lambda_v = 0$, and $\tau_{vt} = 3$, these characteristics for the kinetic energy $\langle E\rangle$ and the Reynolds stress modulus $\langle\sigma_{12}\rangle$ for $\xi = 0$–5 vary in the range $\Delta_F^{\xi} = 0.043$–0.091, and for $\tau_{vt} = 5$, they vary in the range $\Delta_F^{\xi} = 0.017$–0.056. For vibrational relaxation times $\tau_{vt} = 0$–5 and fixed values $\alpha_1 = 0, 5$ and $\xi = 2$, the relative variations of the values of $\langle E(\xi, \tau_{vt})\rangle$ and $\langle\sigma_{12}(\xi, \tau_{vt})\rangle$ are in the range from $\Delta_F^{\tau_{vt}} = 0.103$ for $\tau_{vt} = 0.5$ to $\Delta_F{}^{\tau_{vt}} = 0.01$ for $\tau_{vt} = 5$. In the case where the bulk viscosity reached the maximum value adopted in calculations (see also [10])

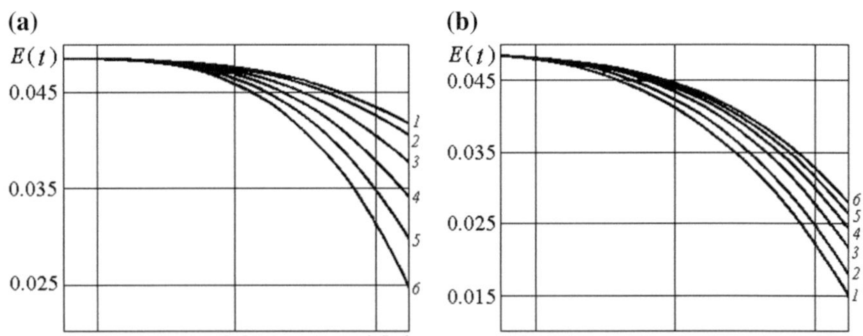

Fig. 6.4 Kinetic energy of perturbation versus time for Re = 100, M = 0.5, Pr = 0.74, $\beta = 0.2$, $\chi = 3, \alpha_1 = 0$–$2, \xi = 0$–$5, \tau_{vt} = 0$–5. **a** is $\eta = \eta_b = \lambda = \lambda_v = 0, \tau_{vt} = 3, \xi = 0.5$ (*1*), 1 (*2*), 2 (*3*), 3 (*4*), 4 (*5*), and 5 (*6*). **b** is $\alpha_1 = 0.5, \xi = 2, \tau_{vt} = 0.5$ (*1*), 1 (*2*), 2 (*3*), 3 (*4*), 4 (*5*), and 5 (*6*)

and corresponding to $\alpha_1 = 2$, for $\xi = 2$, we obtained $\Delta_E^{\tau_{vt}} = 0.158$ for $\tau_{vt} = 3$ and $\Delta_E^{\tau_{vt}} = 0.106$ for $\tau_{vt} = 5$.

It is necessary to note that the ranges of the degree of excitation and relaxation time of the vibrational mode and bulk viscosity corresponded to their real values for nitrogen, oxygen, and carbon monoxide. The results of the numerical modeling suggest that the nonequilibrium of the vibrational mode of the gas molecules has a marked damping effect on the perturbation dynamics at the excitation levels attainable in nozzle flows, underexpanded jets or by moderate laser pumping.

References

1. Boiko, A.V., Grek, G.R., Dovgal, A.V., Kozlov, V.V.: The Origin of Turbulence in Near-Wall Flows. Springer, Berlin (2002)
2. Boiko, A.V., Dovgal, A.V., Grek, G.R., Kozlov, V.V.: Physics of Transitional Shear Flows: Instability and Laminar-Turbulent Transition in Incompressible Near-Wall Shear Layers. Springer, Dordrecht (2012)
3. Browand, F.K., Ho, C.M.: The mixing layer: an example of quasi two-dimensional turbulence. J. Mecanique Teor. Appl. Spec., 99–120 (1983)
4. Abramovich, G.N., Girshovich, T.A., Krasheninnikov, S.Yu., et al.: Theory of Turbulent Jets. Nauka, Moscow (1984) (in Russian)
5. Kovenya, V.M., Yanenko, N.N.: Splitting Method in Problems of Gas Dynamics. Nauka, Novosibirsk (1981) (in Russian)
6. Kochin, N.E., Kibel, I.A., Rose, N.V.: Theoretical Hydromechanics. Part 2. Fizmatgiz, Moscow (1963) (in Russian)
7. Savill, A.M.: Drag reduction by passive devices - a review of some recent developments. In: Gyr, A. (ed.) Structure of Turbulence and Drag Reduction, pp. 429–465. Springer, Berlin (1990)
8. Zhdanov, V.M., Aliyevskii, M.Ya.: Transfer and Relaxation Processes in Molecular Gases. Nauka, Moscow (1989) (in Russian)
9. Nagnibeda, E.A., Kustova, E.V.: Non-equilibrium Reacting Gas Flows. Kinetic Theory of Transport and Relaxation Processes. Springer, Berlin (2009)
10. Grigor'ev, Yu.N., Ershov, I.V., Ershova, E.E.: Influence of vibrational relaxation on the pulsation activity in flows of an excited diatomic gas. J. Appl. Mech. Tech. Phys. **45**, 321–327 (2004)

Chapter 7
Dissipation of the Kelvin–Helmholts Waves in a Relaxing Molecular Gas

Abstract This chapter presents the results of numerical simulations of the full cycle of evolution of the Kelvin -Helmholtz instability, which adequately reproduce the local mechanism of turbulization of the free shear flow. The problem is considered both within the frameworks of the Navier-Stokes equations for a moderate level of thermal nonequilibrium and using the full system of equations of two-temperature aerodynamics for a vibrationally excited gas. Plane waves preliminary calculated by numerical solution of appropriate linearized systems of inviscid gas-dynamic equations are used as initial perturbations. The known pattern of the evolution of the "cat's-eye" large-scale vortex structure typical for the emergence and development of inertial instability is reproduced in detail. The calculated results show the enhancement of dissipation of the kinetic energy of the structure on a background of relaxation process.

The influence of relaxation of excited internal degrees of gas molecules on the evolution of a solitary vortex structure in a linear shear flow was considered in Chap. 6. A drawback of the above-considered model problems is the absence of a mechanism of positive generation of disturbance energy. Moreover, the initial stage of generation of vortex structures by small perturbations enhanced by the kinetic energy of the carrier (mean) flow is not included into this problem. For this reason the effect of thermal nonequilibrium on the entire process of vortex evolution could not be estimated more comprehensively.

In this chapter the dissipative effect of relaxation of excited internal modes of gas molecules on the laminar-turbulent transition process is studied by means of simulating a complete cycle of the development of inertial instability in a time evolving shear layer with an inflection point in the velocity profile. The well-known dynamics of evolution of large vortex structure called a "cat's-eye", which is typical for the emergence and development of the Kelvin–Helmholtz instability, is reproduced in detail. The results of numerical simulations of nonlinear evolution of these disturbances are presented, which providing a sufficiently detailed reproduction of the local

Y.N. Grigoryev and I.V. Ershov, *Stability and Suppression of Turbulence in Relaxing Molecular Gas Flows*, Fluid Mechanics and Its Applications 117,
DOI 10.1007/978-3-319-55360-3_7

mechanism of turbulization of the free shear layer. The problem is considered within the framework of both full Navier–Stokes equations for a moderate level of thermal nonequilibrium and full system of equations of two-temperature aerodynamics for a vibrationally excited gas. The initial perturbations were plane waves preliminary calculated on the basis of appropriately linearized systems of inviscid equations. The calculated results show that the relative enhancement of dissipation of the kinetic energy of the structure averaged over its lifetime can reach 12–15% owing to the increase in the degree of nonequilibrium of internal energy of molecules in realistic ranges for diatomic gases, which agrees well with similar results of the model problem solved in the previous chapter.

7.1 Nonlinear Evolution of the Kelvin–Helmholtz Instability in the Navie–Stokes Model

In this section a development of the Kelvin–Helmholtz instability in free shear layer is studied on the basis of full Navier–Stokes equations, that corresponds to a moderate level of thermal nonequilibrium.

7.1.1 Formulation of the Problem

In the coordinate plane x_1, x_2 we consider a shear flow where the main (carrier) flow of uniform density ρ_0 and temperature T_0 is directed along the abscissa axis x_1 and has the velocity profile

$$U_s(x_2) = U_0 \tanh\left(\frac{x_2}{\delta_0}\right),$$

where δ_0 is the characteristic length determined by the maximum slope of the velocity profile

$$\delta_0 = U_0 \left[\left. \frac{dU_s}{dx_2} \right|_{x_2=0} \right]^{-1} = 1\,.$$

At the time $t = 0$ a two-dimensional perturbation with wavelength λ and wave vector $\mathbf{k} = (\alpha,\ 0)$ is superimposed onto the main flow. The dynamics of the disturbed flow is described by a system of the full Navier–Stokes equations for a compressible viscous heat-conducting gas. Scaling variables are chosen to be the length δ_0, the velocity U_0, density ρ_0, temperature T_0, and the reference time $\tau_0 = \delta_0/U_0$ and pressure $p_0 = \rho_0 U_0^2$ based on these variables. In the dimensionless variables the system of the Navier–Stokes equations has the form

$$\frac{d\rho}{dt} + \rho\left(\frac{\partial u_1}{\partial x_1} + \frac{\partial u_2}{\partial x_2}\right) = 0,$$

$$\rho\frac{du_1}{dt} = -\frac{\partial p}{\partial x_1} + \frac{1}{\mathrm{Re}}\left(\frac{\partial^2 u_1}{\partial x_1^2} + \frac{\partial^2 u_1}{\partial x_2^2}\right) + \frac{(1+3\alpha_1)}{3\mathrm{Re}}\frac{\partial}{\partial x_1}\left(\frac{\partial u_1}{\partial x_1} + \frac{\partial u_2}{\partial x_2}\right),$$

$$\rho\frac{du_2}{dt} = -\frac{\partial p}{\partial x_2} + \frac{1}{\mathrm{Re}}\left(\frac{\partial^2 u_2}{\partial x_1^2} + \frac{\partial^2 u_2}{\partial x_2^2}\right) + \frac{(1+3\alpha_1)}{3\mathrm{Re}}\frac{\partial}{\partial x_2}\left(\frac{\partial u_1}{\partial x_1} + \frac{\partial u_2}{\partial x_2}\right),$$

$$\rho\frac{dT}{dt} + \gamma(\gamma-1)\mathrm{M}^2 p\left(\frac{\partial u_1}{\partial x_1} + \frac{\partial u_2}{\partial x_2}\right) = \frac{\gamma}{\mathrm{Re}\,\mathrm{Pr}}\left(\frac{\partial^2 T}{\partial x_1^2} + \frac{\partial^2 T}{\partial x_2^2}\right)$$

$$+\frac{\gamma(\gamma-1)\mathrm{M}^2}{\mathrm{Re}}\left[\left(\frac{\partial u_1}{\partial x_2} + \frac{\partial u_2}{\partial x_1}\right)^2 + \left(\frac{\partial u_1}{\partial x_1}\right)^2 + \left(\frac{\partial u_2}{\partial x_2}\right)^2\right.$$

$$\left.+\left(\alpha_1 - \frac{2}{3}\right)\left(\frac{\partial u_1}{\partial x_1} + \frac{\partial u_2}{\partial x_2}\right)^2\right],$$

$$\gamma \mathrm{M}^2 p = \rho T, \quad \frac{d}{dt} = \frac{\partial}{\partial t} + u_1\frac{\partial}{\partial x_1} + u_2\frac{\partial}{\partial x_2}. \tag{7.1}$$

The criteria involved in system (7.1) are defined as follows. The Reynolds number is $\mathrm{Re} = U_0\lambda\rho_0/\eta$, the Mach number of the carrier flow is $\mathrm{M} = U_0/\sqrt{\gamma R T_0}$, and the Prandtl number is $\mathrm{Pr} = \eta c_p/\lambda_0$. Here R is the gas constant, $\gamma = c_p/c_v$ is the ratio of specific heats, c_p and c_v are the specific heats at constant pressure and constant volume, respectively, and λ_0 is the thermal conductivity. The coefficient α_1 is equal to the ratio of bulk viscosity to shear viscosity $\alpha_1 = \eta_b/\eta$ and characterizes the degree of nonequilibrium of internal degrees of freedom of gas molecules. It is assumed that the dissipative coefficients in system (7.1) are independent of temperature and have constant values.

The normalized carrier flow is determined by the relations

$$U_s(x_2) = \tanh x_2, \quad T_s = \rho_s = 1, \quad p_s = \frac{1}{\gamma\,\mathrm{M}^2}. \tag{7.2}$$

The boundary-value problem is initially stated in an infinite band whose center coincides with the origin of the coordinate system: $x_1 \in \left[-x_{1,0}; x_{1,0}\right]$, $x_2 \in (-\infty; \infty)$. The band width along the x_1 coordinate is chosen equal to the perturbation wavelength $\lambda = 2\pi/\alpha$. In this case we have $x_{1,0} = \pi/\alpha$. The asymptotic conditions as $x_2 \to \pm\infty$ are changed by $x_2 = \pm x_{2,0}$, where the ordinate $x_{2,0}$ is determined from the condition $\left|\tanh x_{2,0} - 1\right| \le 10^{-9}$. It is assumed in our calculations that $x_{2,0} = 10$.

At $t > 0$ the following periodic conditions are imposed on the boundaries $x_1 = \pm x_{1,0}$, $y \in \left[-x_{2,0}; x_{2,0}\right]$:

$$u_1(t,\, x_{1,0},\, x_2) = u_1(t,\, -x_{1,0},\, x_2), \quad u_2(t,\, x_{1,0},\, x_2) = u_2(t,\, -x_{1,0},\, x_2),$$

$$\rho(t,\, x_{1,0},\, x_2) = \rho(t,\, -x_{1,0},\, x_2), \quad p(t,\, x_{1,0},\, x_2) = p(t,\, -x_{1,0},\, x_2).$$

Correspondingly, at $y = \pm x_{2,0}$, $x \in \left[-x_{1,0};\, x_{1,0}\right]$, the free-stream conditions were imposed:

$$u_1(t,\, x_1,\, x_{2,0}) = -u_1(t,\, x_1,\, -x_{2,0}) = 1, \quad u_2(t,\, x_1,\, x_{2,0}) = u_2(t,\, x_1,\, -x_{2,0}) = 0,$$

$$\rho(t,\, x_1,\, x_{2,0}) = \rho(t,\, x_1,\, -x_{2,0}) = 1, \quad p(t,\, x_1,\, x_{2,0}) = p(t,\, x_1,\, -x_{2,0}) = \frac{1}{\gamma\,\mathrm{M}^2}.$$

The initial conditions for disturbed velocity field and thermodynamic variables are follows

$$u_1(0,\, x_1,\, x_2) = \tanh x_2 + u'_1(0,\, x_1,\, x_2), \quad u_2(0,\, x_1,\, x_2) = u'_2(0,\, x_1,\, x_2),$$

$$\rho(0,\, x_1,\, x_2) = 1 + \rho'(0,\, x_1,\, x_2), \quad T(0,\, x_1,\, x_2) = 1 + T'(0,\, x_1,\, x_2),$$

$$p(0,\, x_1,\, x_2) = \frac{1}{\gamma\mathrm{M}^2} + p'(0,\, x_1,\, x_2). \tag{7.3}$$

The perturbations of the velocity vector components u'_1, u'_2 and of the thermodynamic properties ρ', T', and p' are initially calculated on the basis of the results obtained in [1].

7.1.2 *Calculation of Initial Perturbations*

The initial perturbations superimposed onto the main flow are chosen to be the parameters of natural linear inviscid oscillations with the most growth rates, which were considered in Chap. 2. They are determined as solutions of the system of gas-dynamic equations linearized with respect to the steady-state main flow (7.2):

$$\frac{\partial \rho'}{\partial t} + U_s\frac{\partial \rho'}{\partial x_1} = -\left(\frac{\partial u'_1}{\partial x_1} + \frac{\partial u'_2}{\partial x_2}\right), \quad \frac{\partial u'_1}{\partial t} + U_s\frac{\partial u'_1}{\partial x_1} + u'_2\frac{\partial U_s}{\partial x_2} = -\frac{\partial p'}{\partial x_1},$$

$$\frac{\partial u'_2}{\partial t} + U_s\frac{\partial u'_2}{\partial x_1} = -\frac{\partial p'}{\partial x_2}, \quad \mathrm{M}^2\left(\frac{\partial p'}{\partial t} + U_s\frac{\partial p'}{\partial x_1}\right) + \frac{\partial u'_1}{\partial x_1} + \frac{\partial u'_2}{\partial x_2} = 0,$$

$$p' = \frac{1}{\gamma\,\mathrm{M}^2}\left(\rho' + T'\right). \tag{7.4}$$

Table 7.1 Wavenumbers and phase velocities of the most growing inviscid disturbances

M	α	c_i	$\alpha\, c_i$
0.0	0.445	0.427	0.190
0.2	0.426	0.426	0.181
0.5	0.397	0.356	0.141
0.7	0.326	0.309	0.101

Here the equations for disturbances of density and static temperature are changed by the equation for pressure disturbance (see Chap. 2). The sought solutions are presented as plane waves propagating in the main flow direction

$$\left\{u_1', u_2', \rho', T', p'\right\}$$
$$= \left\{u(x_2), v(x_2), \widehat{\rho}(x_2), \widehat{T}(x_2), \widehat{p}(x_2)\right\} \exp\left[i\alpha\left(x_1 - ct\right)\right], \qquad (7.5)$$

where u, v, $\widehat{\rho}$, $\widehat{T}$, and $\widehat{p}$ are the amplitude functions of disturbances, i is the imaginary unity, and $c = c_r + ic_i$ is the complex phase velocity of the wave.

Substitution of Eq. (7.5) into system (7.4) yields a spectral problem, which is transformed at $c_r = 0$ to a scalar equation for $\widehat{p}$:

$$\frac{d^2\widehat{p}}{dx_2^2} - \frac{2\,\mathrm{sech}^2\, x_2}{(\tanh\, x_2 - ic_i)}\frac{d\widehat{p}}{dx_2} - \alpha^2\left[1 - \mathrm{M}^2\left(\tanh\, x_2 - ic_i\right)^2\right]\widehat{p} = 0,$$

$$\widehat{p}\big|_{x_2=-\infty} = \widehat{p}\big|_{x_2=+\infty} = 0. \qquad (7.6)$$

Here $\widehat{p} = \widehat{p}_r + i\widehat{p}_i$, $\widehat{p}_r(x_2)$ and $\widehat{p}_i(x_2)$ are the real and imaginary parts of the complex eigenfunction of pressure, and eigenvalues of problem (7.6) are the wavenumbers α.

Based on Eq. (7.6) the wavenumbers α, phase velocities c_i, and growth rates $\alpha\, c_i$ of the most growing inviscid modes were found for a subsonic carrier flow with $\mathrm{M} < 1$. Their values for various Mach numbers used in our calculations are presented in Table 7.1.

Besides the spectral parameters for statement of initial date the eigenfunctions of the spectral problem are needed, which are calculated in this work by means of numerical integration of problem (7.6) for the values of the parameters from Table 7.1. Equation (7.6) is divided into the real and imaginary parts, which are converted to the normal system of the first-order equations. The Cauchy data at $x_2 = 0$ for this problem are chosen similar to [1]. The system is integrated on the interval $\left[-x_{2,0};\; x_{2,0}\right]$ by the fourth-order Runge–Kutta procedure. Other amplitude functions are calculated via the found functions $\widehat{p}$ with the use of relations that follow from Eqs. (7.4) and (7.5):

$$u = \frac{i}{\alpha}\left(\frac{\tanh\, x_2 + ic_i}{\tanh^2\, x_2 + c_i^2}\right)\left(v\,\mathrm{sech}^2 x_2 + i\alpha\hat{p}\right), \quad v = \frac{i}{\alpha}\left(\frac{\tanh\, x_2 + ic_i}{\tanh^2\, x_2 + c_i^2}\right)\frac{d\widehat{p}}{dx_2},$$

$$\widehat{\rho} = \mathrm{M}^2\widehat{p}, \qquad \widehat{T} = (\gamma - 1)\mathrm{M}^2\widehat{p}.$$

As a result, the initial perturbations (7.3) introduced into the calculations are determined as the real part of the calculated solutions (7.5)

$$\mathbf{q}'(x_1, x_2, t) = Real\Big[\widehat{\mathbf{q}}(x_2)\,\exp\big(i\alpha x_1\big)\Big]\,\exp\big(\alpha c_i t\big).$$

where

$$\mathbf{q}' = \big(u'_1, u'_2, \rho', T', p'\big), \quad \widehat{\mathbf{q}}(x_2) = \big(u, v, \widehat{\rho}, \widehat{T}, \widehat{p}\big),$$

It should be noted that smaller wave numbers (greater Mach numbers) correspond to greater amplitudes of superimposed perturbations.

7.1.3 Numerical Calculations of the Evolution of Perturbations

The calculations are performed on a regular grid with an identical step h over both spatial variables. System (7.1) is approximated by a weighted finite-difference scheme with splitting over the physical processes and spatial variables [2], which was already used in previous Chap. 6. In the operator form the scheme is written as follows [2, 3]:

$$\frac{(\mathbf{f}_h^{n+1} - \mathbf{f}_h^n)}{\tau} + \mathbf{A}_h^{(2)}[\delta\mathbf{f}_h^{n+1} + (1-\delta)\mathbf{f}_h^n] = \mathbf{R}_h^n. \tag{7.7}$$

Here $\mathbf{f}_h^n = (\rho_h^n, u_{1,h}^n, u_{2,h}^n, T_h^n)$ is the grid vector-function of the solution at the nth time layer in the (i, j)th node, τ is the time step, and δ is the weight parameter. The operator $\mathbf{A}_h^{(2)}$ includes symmetric approximations with the second order of the first and second spatial derivatives on a three-point mould over each spatial coordinate. The operator $\mathbf{R}_h^n$ in the right side of equality (7.7) consists of symmetric approximations over each spatial variable with the second order of mixed derivatives from the momentum equation and terms of the dissipative function from the energy equation. The thus-defined difference scheme (7.7) approximates the initial differential system (7.1) with the order $O(\tau + h^2)$ and is absolutely stable at $\delta \geq 0.5$.

In accordance to [2] scheme (7.7), after approximate factorization, is implemented by the method of fractional steps. The cyclic sweep method is applied over the periodic coordinate x_1; the algorithm of its implementation is taken from [4]. The steps $h = 0.1$ and $\tau = 0.01$ in time and space, respectively, are used in calculations.

7.1.4 Effect of Bulk Viscosity on Vorticity Kinematics

All computations are performed for M = 0.2–0.5 and Re = 100. The value of the Prandtl number Pr = 0.74, the ratios $\gamma = 1.4$, and $\alpha_1 = 0$–2 correspond to diatomic gases. For these values of the parameters the dissipative effect of bulk viscosity was considered in a model problem in Chap. 6.

The evolution of a vortex perturbation in a compressible unsteady flow is considered on the basis of the behavior of the field of the total vorticity isolines. The grid function of the total vorticity is calculated by using approximations with the second order of the transverse component of the vorticity vector by the central differences

$$\omega_{ij}^{n} = \frac{u_{2,\,i+1,j} - u_{2,\,i-1,j}}{2h} - \frac{u_{1,\,i,j+1} - u_{1,\,i,j-1}}{2h},$$

where $u_{1,\,i,j}$ and $u_{2,\,i,j}$ are the grid functions of the components of the flow velocity vector, which are calculated on the basis of the difference approximation (7.7).

The calculations reproduce the details of the known pattern of the large eddy dynamics, which is typical for the emergence and development of the Kelvin–Helmholtz instability [5–7]. The isolines of the vorticity field ω at the initial time $t = 0$ are shown in Fig. 7.1 for the Mach numbers M = 0.2 and 0.5 and the Reynolds number Re = 100. It is seen that the perturbation inserted into the flow is concentrated at the beginning in a narrow band of low velocities in the vicinity of $x_2 = 0$ where closed isolines identifying a pair of vortices of identical intensity are formed. Outside this band $x_2 = \pm 2$, however, the perturbation is vanishingly small. From comparison of Fig. 7.1a, b, the initial perturbations increase with increasing Mach number owing to the influence of compressibility.

At later times the middle part of the disturbed region rapidly expands, the vortices merge with each other, and a close-to-circular subdomain of concentrated vorticity with closed isoline (the so-called core) is formed. Simultaneously, comparatively narrow layers (braids) are formed on both sides of the core, which serve for transportation of vorticity from the periphery to the core. In turn the core generates tensile stresses in the braids, which help to maintain them. As the system is periodic in space, there arises a "cat's-eye" structure, where the geometry of isolines in the cores is topologically equivalent to singular points of the "center" type, whereas the geometry of isolines on the interfaces is topologically equivalent to "saddle"-type singular points [8]. The structure rapidly reaches the maximum size at $t \simeq 2$, after which the disturbance decays. A typical picture of vorticity isolines close to the maximum size of the structure is shown in Fig. 7.2. It can be noted that the core radius at this moment is 1.5 times greater and the width of the layer on the boundary is 1.5 times smaller than the transverse size of the initial perturbed region in Fig. 7.1. The disturbance decays until the moment $t \simeq 3.5$, after which the structure is stabilized.

It is known [5] that the mechanism of the development of the Kelvin–Helmholtz instability has an inviscid inertial character, and the dynamic viscosity (Reynolds number) produces a comparatively small effect on the kinematics of large eddy

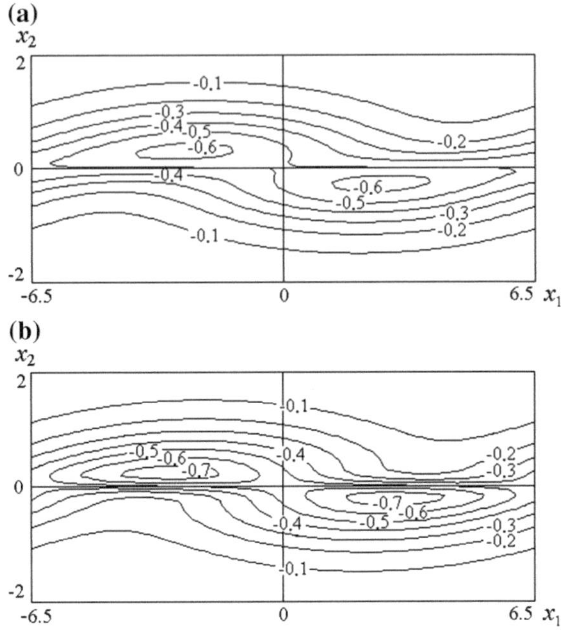

Fig. 7.1 Vorticity field isolines ω at the time $t = 0$ for Re = 100. **a** is M = 0.2. **b** is M = 0.5

evolution [6, 7]. For this reason it could be expected that the additional effect of bulk viscosity will be even less pronounced.

The degree of bulk viscosity influence on the vorticity kinematics is estimated on the basis of the time dependence of a certain conventional area of the vortex structure for different values of the coefficient α_1. The boundary of this area is taken to be the first nonclosed vorticity isoline incoming into the singular point ("saddle") on the boundary. It corresponds to the vorticity value $\omega = -0.15$ (see Fig. 7.2).

To find the time dependence of the conventional area of the structure $S(t,\ \alpha_1)$, an additional regular grid with the steps

$$\Delta x_{1,\,i} = \Delta x_{2,\,j} = h = 0.05$$

is introduced in the computational domain, so that the centers of the cells of this grid coincide with the nodes of the initial grid used to calculate the fields of the hydrodynamic variables. The vorticity values at the additional nodes are found by means of linear interpolation. The sought area is determined as

$$S(t,\ \alpha_1) = \sum_{i=1}^{N_1} \sum_{j=1}^{N_2} \left(\Delta x_{1,\,i}\, \Delta x_{2,\,j}\, \zeta_{ij} \right),$$

where ζ_{ij} has the form

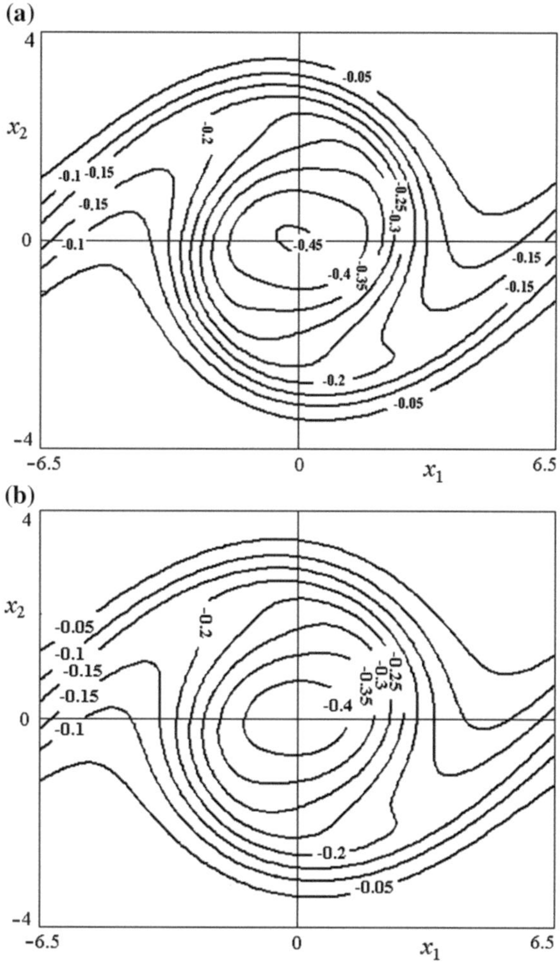

Fig. 7.2 Vorticity field isolines ω at the time $t = 2.5$ for Re $= 100$, M $= 0.5$. **a** is $\alpha_1 = 0$. **b** is $\alpha_1 = 2$

$$\zeta_{ij} = \begin{cases} 1, & \omega_{ij} \leq -0.15, \\ 0, & \omega_{ij} > -0.15. \end{cases}$$

Figure 7.3 shows the graphic dependencies of the variable

$$\mu(t, \alpha_1) = \frac{S(t,\ \alpha_1)}{S(t,\ 0)}$$

for the Reynolds number Re $= 100$, Mach numbers M $= 0.2$ and 0.5, and different values of the parameter α_1. It is seen that stratification of the curves along the parameter α_1, which is associated with the additional dissipative effect of bulk viscosity is not very significant. For all values of this parameter the structure evolution has a universal character: growth and reaching the maximum value at $t \simeq 2$, followed by

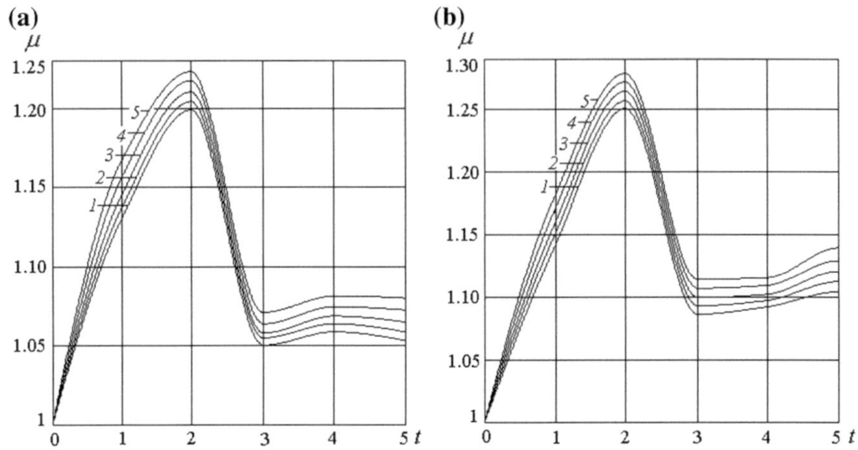

Fig. 7.3 Conventional area $\eta(t, \alpha_1)$ versus time for Re $= 100$, $\alpha_1 = 0$ (*1*), 0.5 (*2*), 1 (*3*), 1.5 (*4*), and 2 (*5*). **a** is M $= 0.2$. **b** is M $= 0.5$

Table 7.2 Normalized deviations $\varepsilon_S(\alpha_1)$, % for Re $= 100$

α_1	0.0	0.5	1.0	1.5	2.0
M $= 0.2$	0.0	0.468	0.953	1.595	2.253
M $= 0.5$	0.0	0.611	1.239	1.952	2.686

reduction and stabilization at $t > 5$ to a certain value slightly greater than the initial value. Comparing Fig. 7.3a, b, we can see that the greatest values at the nonlinear stage of evolution are observed for disturbances with a greater wavelength (greater value of M), in contrast to the linear theory [1], which predicts that disturbances with short wave lengths reach the maximum values (see Table 7.1). It may be assumed that this effect, which was noted for an incompressible stratified fluid in [6], is enhanced by the influence of compressibility, as it takes place for the initial perturbations.

To quantify the influence of bulk viscosity (parameter α_1) on smearing of the vortex structure, we calculated the time-averaged normalized deviations

$$\varepsilon_S(\alpha_1) = \Theta^{-1} \int_0^{\Theta} |\mu(t, \alpha_1) - 1| \; 100\% \; dt, \quad \alpha_1 = 0\text{--}2.$$

The calculated normalized deviations $\varepsilon_S(\alpha_1)$ for the Mach numbers M $= 0.2$ and 0.5, Reynolds number Re $= 100$, and interval of time averaging $\Theta = 6$ are listed in Table 7.2.

It follows from Table 7.2 that an increase in the parameter α_1 for a fixed Mach number leads to more intense spreading of the vortex structure in space, though the quantitative manifestation of this effect is very small, as it could be expected. It is

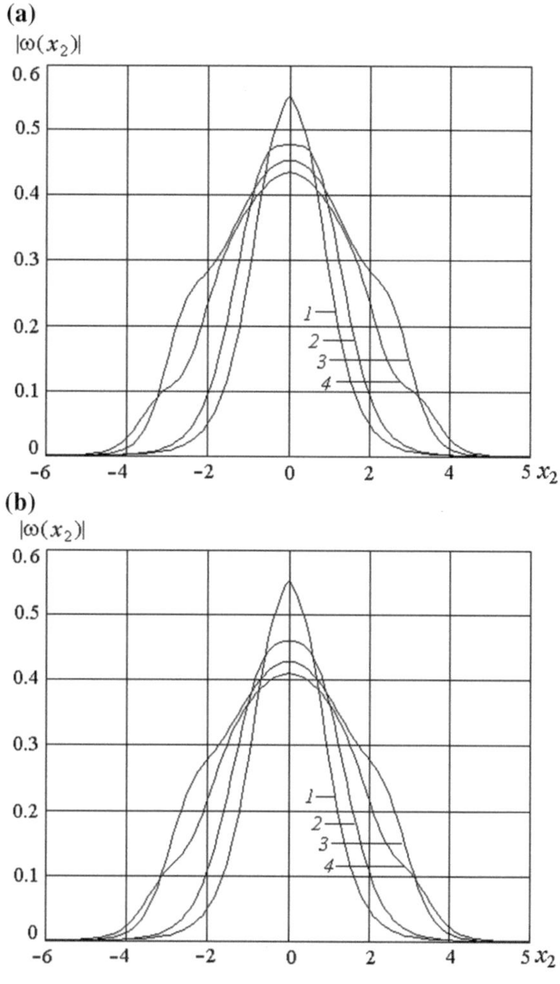

Fig. 7.4 Dependencies $|\,\omega(0, x_2)|$ for fixed time moments t for Re $= 100$, M $= 0.5$, and $t = 0$ (*1*), 1 (*2*), 2.5 (*3*), and 4 (*4*). **a** is $\alpha_1 = 0$. **b** is $\alpha_1 = 2$

more interesting to note that the level of spreading at a fixed value of α_1 is enhanced approximately in proportion to the increase in the Mach number. However, the narrow range of the Mach number M in the calculations performed does not allow us to trace the effect of compressibility on the vorticity kinematics in more detail.

The quantitative pattern of the time evolution of the vorticity is illustrated in Fig. 7.4, which shows the profiles $|\omega|$ in the cross section $x_1 = 0$ passing through the center of the structure. It is known [8] that the total circulation of a localized vortex structure is retained in time within the framework of the Navier–Stokes equations. Therefore, vortex spreading can be evaluated on the basis of vorticity reduction at the vortex center. In this case the center of the structure coincides with the origin of the coordinate system, which is identically fixed in the computational grid, in contrast to the dividing isoline bounding the area $S(t,\ \alpha_1)$.

Table 7.3 Normalized deviations $\varepsilon_\omega(\alpha_1)$, % for Re = 100

α_1	0.0	0.5	1.0	1.5	2.0
M = 0.2	0.0	0.663	1.546	2.592	3.679
M = 0.5	0.0	0.949	2.216	3.704	5.227

To estimate the influence of the parameter α_1 on the absolute value of the vorticity at the center of the structure $|\omega(0,\, t,\, \alpha_1)| = \omega_0(t,\, \alpha_1)$ we calculate the normalized deviations

$$\varepsilon_\omega(\alpha_1) = \Theta^{-1} \int\limits_0^{\Theta} \left| \frac{\omega_0(t,\, \alpha_1) - \omega_0(t,\, 0)}{\omega_0(t,\, 0)} \right| 100\%\, dt, \quad \alpha_1 = 0\text{–}2,$$

where the time-averaging interval $\Theta = 6$ can be considered as the "lifetime" of the structure. The calculated normalized deviations $\varepsilon_\omega(\alpha_1)$ for the Reynolds number Re = 100 and Mach numbers M = 0.2 and 0.5 are summarized in Table 7.3.

We can state that the data in Table 7.3 more definitely confirm the conclusions about the character of the influence of the parameter α_1 and the Mach number on the vortex structure kinematics, which were drawn on the basis of the results in Table 7.2.

7.1.5 Dissipation of the Kinetic Energy of Disturbances

The most important effect from the physical viewpoint is the influence of bulk viscosity on the energy exchange between the carrier flow and disturbances, which are reproduced more realistically in the present problem than in the model problem considered in Chap. 6. Similar to [1, 3], we study the time evolution of the kinetic energy of disturbance

$$E(t) = \frac{1}{2} \int\limits_{-x_{1,0}}^{x_{1,0}} dx_1 \int\limits_{-x_{2,0}}^{x_{2,0}} dx_2\ \rho\ \left(u_1'^2 + u_2'^2\right) \tag{7.8}$$

and the absolute value of the Reynolds stresses

$$\sigma_{12}(t) = \int\limits_{-x_{1,0}}^{x_{1,0}} dx_1 \int\limits_{-x_{2,0}}^{x_{2,0}} dx_2\ \left|\rho\, u_1' u_2'\right|. \tag{7.9}$$

The value of the disturbance energy production is calculated on the basis of the integral energy balance equation derived in [9]

$$D(t) = \frac{dE}{dt} = J_1 + J_2 - \frac{1}{\mathrm{Re}}\,(J_3 + \alpha_1 J_4)\,, \tag{7.10}$$

where the term

$$J_1 = -\int\limits_{-x_{1,0}}^{x_{1,0}} dx_1 \int\limits_{-x_{2,0}}^{x_{2,0}} dx_2\, \rho\, u'_1 u'_2\, \frac{dU_s}{dx_2} \tag{7.11}$$

describes the energy exchange between the disturbance and the main flow. The integral

$$J_2 = \int\limits_{-x_{1,0}}^{x_{1,0}} dx_1 \int\limits_{-x_{2,0}}^{x_{2,0}} dx_2\, p' \left(\frac{\partial u'_1}{\partial x_1} + \frac{\partial u'_2}{\partial x_2}\right) \tag{7.12}$$

is interpreted as the work performed during pulsed expansion (compression) of the gas. The integrals

$$J_3 = \int\limits_{-x_{1,0}}^{x_{1,0}} dx_1 \int\limits_{-x_{2,0}}^{x_{2,0}} dx_2 \left[\left(\frac{\partial u'_1}{\partial x_1}\right)^2 + \left(\frac{\partial u'_1}{\partial x_2}\right)^2 + \left(\frac{\partial u'_2}{\partial x_1}\right)^2 + \left(\frac{\partial u'_2}{\partial x_2}\right)^2 + \frac{1}{3}\left(\frac{\partial u'_1}{\partial x_1} + \frac{\partial u'_2}{\partial x_2}\right)^2\right], \tag{7.13}$$

$$J_4 = \int\limits_{-x_{1,0}}^{x_{1,0}} dx_1 \int\limits_{-x_{2,0}}^{x_{2,0}} dx_2 \left(\frac{\partial u'_1}{\partial x_1} + \frac{\partial u'_2}{\partial x_2}\right)^2 \tag{7.14}$$

determine the contributions of the dissipative processes associated with shear and bulk viscosity, respectively.

In Eqs. (7.11)–(7.14), u'_1, u'_2, and p' are the fluctuations of the velocity components and pressure, $x_{1,0} = \pi/\alpha$, $x_{2,0} = 10$, and the wave numbers α for different Mach numbers M are listed in Table 7.1. The integrals are calculated by quadrature formulas of trapezoids with a step $h = 0.1$ on the grid used in our computations. The fluctuating characteristics of the flow are determined as the differences

$$\mathbf{F}'(t) = \mathbf{F}(t) - \mathbf{F}_s(t),$$

where the components of the vector-function $\mathbf{F}(t)$ are instantaneous values of the disturbed flow characteristics, and $\mathbf{F}_s(t)$ are the corresponding characteristics of the carrier flow. Because the undisturbed flow in this case, in contrast to the model problem of Chap. 6, is not an exact steady-state solution of the Navier–Stokes equations (7.1), its instantaneous characteristics are calculated simultaneously with the disturbed flow calculation.

To verify the calculation of the energy production $D(t)$ from the energy balance equation (7.10), the values of $D(t)$ are calculated directly by Eq. (7.8) with using symmetric finite difference approximation of the time derivative

$$D(t) = \frac{E(t+\tau) - E(t-\tau)}{2\tau},$$

The alternative values of $D(t)$ differ by less than 1.5%.

The time evolution of the kinetic energy of the disturbances $E(t,\ \alpha_1)$ is illustrated in Fig. 7.5. The greater value of the parameter α_1, the smaller the kinetic energy of the structure in the entire time interval. The evolution of $E(t,\ \alpha_1)$ conforms to the kinematics of the structure evolution, which was contemplated in the comment to Fig. 7.2. As the structure size increases, the fluctuating energy also increases, until it reaches the maximum value and then decreases and stabilizes to a certain value at the time $\Theta > 6$. The shift of the peaks in the plots of Figs. 7.2 and 7.5 is caused by the conventional calculation of the structure area $S(t,\ \alpha_1)$, whereas the energy $E(t,\ \alpha_1)$ is calculated by means of integration over the entire computational domain. The time dependencies of the absolute value of the Reynolds stresses $\sigma_{12}(t,\ \alpha_1)$ determined by Eq. (7.9) almost completely repeat the plots of $E(t,\ \alpha_1)$.

For quantitative comparison of the contributions of bulk viscosity to the dissipation of the kinetic energy of the disturbances, the time-averaged normalized deviations are calculated as

$$\varepsilon_E(\alpha) = \Theta^{-1} \int\limits_0^{\Theta} \left| \frac{E(t,\ \alpha_1) - E(t,\ 0)}{E(t,\ 0)} \right| 100\%\, dt, \quad \alpha_1 = 0\text{–}2.$$

The averaging is performed over a conventional "lifetime" of the structure $\Theta = 6$. The calculated normalized deviations $\varepsilon_E(\alpha_1)$ are listed in Table 7.4.

It follows from Table 7.4 that an increase in the parameter α for a fixed Mach number leads to more intense dissipation of the kinetic energy of the disturbance. At the Mach number $\mathrm{M} = 0.5$ the time-averaged normalized decrease in energy for $\alpha_1 = 2$ reaches $\varepsilon_E = 13.567\%$. This value agrees well with similar data obtained in the Chap. 6 for the model structure, though that study included only the decay of the structure.

The calculated dependencies of the fluctuating energy production $D(t)$, and also individual contributions J_1, J_3, and J_4 to the overall balance of the process obtained from the energy equation (7.10) are shown in Figs. 7.6, 7.7, 7.8 and 7.9. Analysis of the plots in Figs. 7.6, 7.7, 7.8 and 7.9 allows one to find more details of the behavior of the kinetic energy of the structure and its dissipation. In the time interval where the structure size and the kinetic energy of the disturbance increase the energy production is positive at first and increases, until it reaches a certain peak value, after which it starts to decrease; finally, the energy production becomes negative, which corresponds to a decrease in the energy of the structure. There is complete correspondence of the characteristic points in Figs. 7.5 and 7.6. In particular the

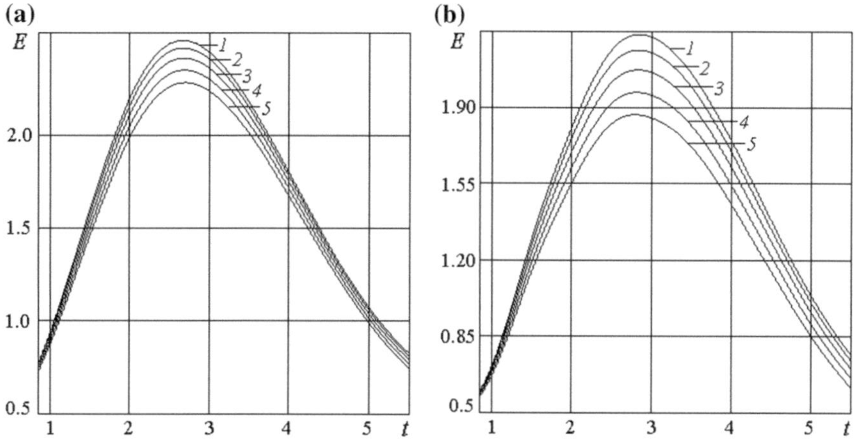

Fig. 7.5 Time evolution of the disturbance energy $E(t)$ for Re = 100, $\alpha_1 = 0$ (*1*), 0.5 (*2*), 1 (*3*), 1.5 (*4*), and 2 (*5*). **a** is M = 0.2. **b** is M = 0.5

Table 7.4 Normalized deviations $\varepsilon_E(\alpha_1)$, % for Re = 100

α_1	0.0	0.5	1.0	1.5	2.0
M = 0.2	0.0	1.347	3.148	5.287	7.511
M = 0.5	0.0	2.478	5.772	9.633	13.567

points where the curves $D(t)$ pass through zero correspond to the maximum points on the curves $E(t)$, whereas the maximum and minimum points of the curve $D(t)$ correspond to the inflection points on the curves $E(t)$.

A comparison of the plots in Figs. 7.6 and 7.7 shows that the disturbance energy production is almost completely determined by the contribution of the integral $J_1(t)$, which characterizes the energy exchange between the disturbance and the carrier flow in Eq. (7.10). In the integrand of $J_1(t)$ the derivative of the carrier flow velocity is greater than zero ($dU_s/dx_2 > 0$) in the entire calculated time interval. Simultaneously, at the first stage of the structure evolution, where $J_1(t) > 0$, the Reynolds stresses $\rho u'_1 u'_2$ are also at least nonnegative in the major part of the computational domain. Energy pumping to the vortex structure from the external flow occurs, which is responsible for the increase in its kinetic energy. Such a direction of the energy flux in the phenomenological theory of turbulence is usually associated with the "negative" viscosity phenomenon. The transition of $J_1(t)$ to the negative domain is determined by the change in the sign of the Reynolds stresses, which become negative in the major part of the domain and induce an usual dissipative effect, thus, assisting in the disturbance decay.

Comparing the plots in Figs. 7.6, 7.7, 7.8 and 7.9, we may conclude that the negative contribution of the dissipative terms $J_3(t)$ and $J_4(t)$ produces a more appreciable

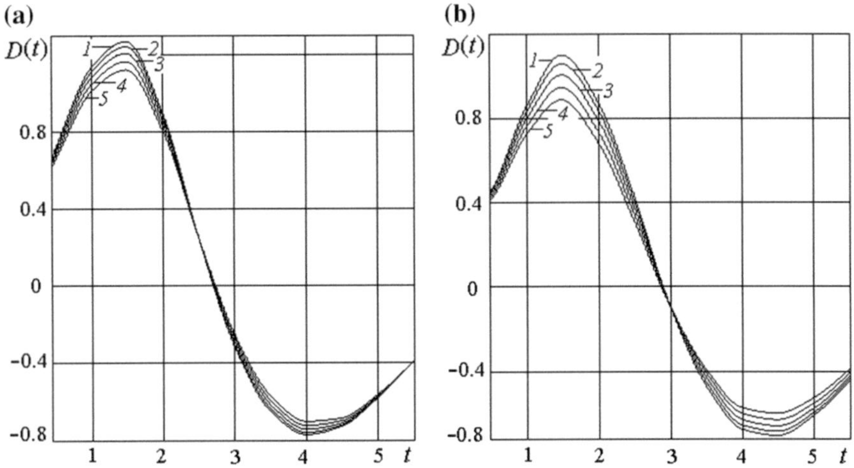

Fig. 7.6 Fluctuating energy production $D(t)$ versus time for Re = 100, $\alpha_1 = 0$ (*1*), 0.5 (*2*), 1 (*3*), 1.5 (*4*), and 2 (*5*). **a** is M = 0.2. **b** is M = 0.5

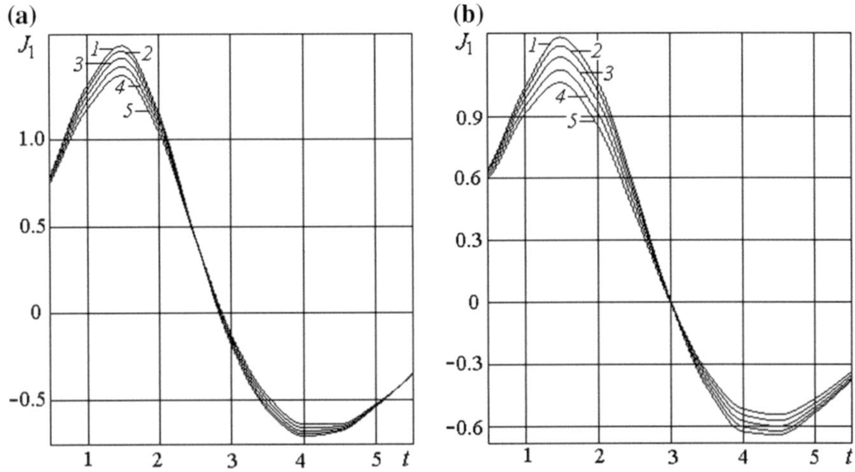

Fig. 7.7 Integral $J_1(t)$ versus time for Re = 100, $\alpha_1 = 0$ (*1*), 0.5 (*2*), 1 (*3*), 1.5 (*4*), and 2 (*5*). **a** is M = 0.2. **b** is M = 0.5

effect on the production of the kinetic energy of the disturbance at times close to the positive maximums of $D(t)$ and $J_1(t)$, that can be observed by comparing Figs. 7.6 and 7.7. At the same time, we can argue that the zero points of these graphs have a minor shift in time, whereas the negative production in the domain $D(t) < 0$ is mainly associated with the behavior of $J_1(t) < 0$. In particular stratification of both dependencies along the parameter α_1 has the same character: greater values of bulk viscosity correspond to smaller decay rates $D(t)$ (less negative values of $J_1(t)$). Thus, we observe the opposite behavior to that obtained in Chap. 6, where the decay rate

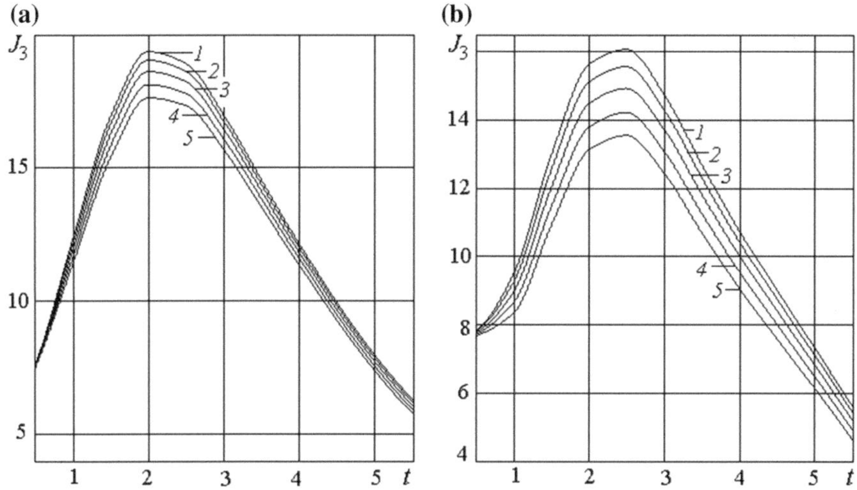

Fig. 7.8 Integral $J_3(t)$ versus time for Re $= 100$, $\alpha_1 = 0$ (*1*), 0.5 (*2*), 1 (*3*), 1.5 (*4*), and 2 (*5*). **a** is M $= 0.2$. **b** is M $= 0.5$

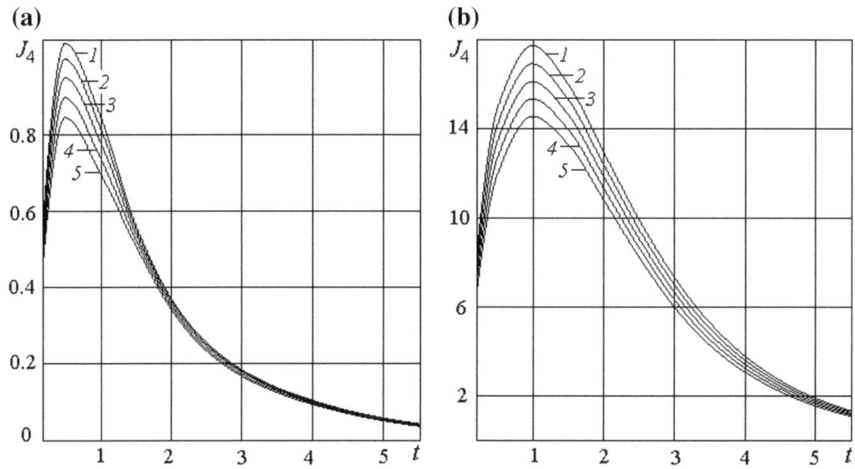

Fig. 7.9 Integral $J_4(t)$ versus time for Re $= 100$, $\alpha_1 = 0$ (*1*), 0.5 (*2*), 1 (*3*), 1.5 (*4*), and 2 (*5*). **a** is M $= 0.2$. **b** is M $= 0.5$

increased in proportion to the increase in α_1. This means that the damping of the kinetic energy of the disturbance at $D(t) < 0$ is not determined by negative contributions of $J_3(t)$ and $J_4(t)$. Therefore, the Reynolds number Re $= 100$ is appreciably greater than the critical Reynolds number, which is found by solving the variational problem of the energy theory of stability obtained in Chap. 5 from Eq. (7.10) for $D(t) = 0$.

Table 7.5 Time evolution of the integral $J_2(\alpha_1)$ for Re = 100 and M = 0.5

α_1	0.0	0.5	1.0	1.5	2.0
$t = 1$	0.036219	0.036590	0.036144	0.035265	0.034306
$t = 2$	−0.039891	−0.037281	−0.034514	−0.031827	−0.029524
$t = 3$	0.000170	0.000300	0.000270	0.000213	0.000197
$t = 4$	0.021402	0.020719	0.020008	0.019332	0.018742
$t = 5$	−0.005725	−0.005292	−0.005101	−0.005091	−0.005002

A comparison of the values of the integrals $J_3(t)$ and $J_4(t)$ plotted in Figs. 7.8 and 7.9 shows that the dissipative contribution of the shear viscosity at $\mathrm{M} = 0.2$ is greater than the corresponding contribution of bulk viscosity by an order of magnitude. However, already for $\mathrm{M} = 0.5$, both contributions become comparable ($J_3(t) \sim J_4(t)$), especially in view of $\alpha_1 > 1$. This allows us to assume that the dissipative effect associated with the relaxation process may become a key factor of disturbance suppression in the case of more intense excitation involving the vibrational degrees of freedom, as was obtained in the model problem of Chap. 6.

Some selected values of the integral $J_2(t, \alpha_1)$, which describe the processes of oscillatory expansion and compression in Eq. (7.10) at different time moments t for $\mathrm{M} = 0.5$ and $\mathrm{Re} = 100$ are listed in Table 7.5.

Comparing the data in Table 7.5 with the characteristic values of the other energy balance integrals J_1, J_3, and J_4, we can see that they differ by more than two orders of magnitude, and the contribution of $J_2(t)$ in the first approximation is negligibly small. Considering the integrands in J_2 and J_4, we can see that a possible reason could be the small values of the pressure fluctuations p', which are related via the equation of state (see system (7.1)) to the temperature fluctuations T' and density fluctuations ρ':

$$p' = \frac{1}{\gamma \mathrm{M}^2}\left(\rho T' + \rho' T_s\right).$$

In Chap. 5 the energy balance equation was written with respect to the production of the total energy of the disturbances, which was expressed by the integral

$$E_t(t) = \int_{\Omega} \left\{ \frac{\rho_s u_i'^2}{2} + \frac{1}{2\gamma \mathrm{M}^2}\left[\left(\frac{\rho'}{\rho_s}\right)^2 + \frac{1}{(\gamma-1)}\left(\frac{\rho T'^2}{2T_s}\right)\right]\right\} d\Omega.$$

The additional terms here, as compared to the integral of the kinetic energy $E(t)$ in Eq. (7.10), are obtained by the transformation of the integral $J_2(t)$. The comparison of the magnitude of the individual contributions shows that it is possible to neglect the integral $J_2(t)$ in the equation of the form of Eq. (7.10) for a certain class of compressible flows in the energy theory of stability and to consider only the production of the kinetic energy of the disturbances.

The results of the study allow us to evaluate quantitatively bulk viscosity influence on the formation and evolution of the naturally developing vortex structure and to formulate the following remarks.

The influence of bulk viscosity on the kinematics of the vortex structure with a predominantly inviscid mechanism of evolution is comparatively weak, but its additional dissipative effect is clearly manifested.

As bulk viscosity increases within the realistic limits for diatomic gases, the normalized enhancement of the dissipation of the kinetic energy of the structure averaged over its "lifetime" reaches 13%, which agrees well with similar results obtained for the model problem of Chap. 6.

In the subsonic range of Mach numbers (M $\geq$ 0.5), the contributions of bulk viscosity (compressibility) and shear viscosity to the kinetic energy dissipation rate become comparable in the order of magnitude.

For disturbed plane-parallel subsonic flows of molecular gases the disturbance energy evolution is sufficiently accurately described by the simple functional of the kinetic energy of the disturbances.

7.2 Effect of a Vibrational Relaxation on the Kelvin–Helmholtz Instability

In this section the influence of the vibrational relaxation on suppression of the Kelvin–Helmholtz instability in an evolving shear layer is studied numerically on the basis of equations of two-temperature aerohydrodynamics. Planar waves with the maximum growth rates, which were computed in Chap. 2 within the framework of a linearized system of equations of inviscid two-temperature gas dynamics, are used as the initial disturbances.

7.2.1 Formulation of Problem

A plane-parallel shear flow of a vibrationally nonequilibrium diatomic gas in a coordinate plane (x_1, x_2) is considered. The stationary (carrier) flow with a uniform thermodynamic parameters: density ρ_0 and temperatures $T_0 = T_{v,0}$ is directed along the abscissa axis x_1 and has a velocity profile with an inflection point at $x_2 = 0$.

The characteristic quantities for normalization are chosen as in (7.1). In the dimensionless variables the system of equations of two-temperature aerodynamics has the form

$$\frac{\partial \rho}{\partial t} + \frac{\partial \rho u_i}{\partial x_i} = 0,$$

$$\rho\left(\frac{\partial u_i}{\partial t} + u_j\frac{\partial u_i}{\partial x_j}\right) = -\frac{\partial p}{\partial x_i} + \frac{1}{\text{Re}}\frac{\partial}{\partial x_j}\left[\eta(T)\frac{\partial u_i}{\partial x_j}\right] + \frac{1}{\text{Re}}\left(\alpha_1 + \frac{1}{3}\right)\frac{\partial}{\partial x_i}\left[\eta(T)\frac{\partial u_j}{\partial x_j}\right],$$

$$\rho\left(\frac{\partial T}{\partial t}+u_j\frac{\partial T}{\partial x_j}\right)+(\gamma-1)\rho T\frac{\partial u_i}{\partial x_i}=\frac{\gamma}{\mathrm{Re}\,\mathrm{Pr}}\frac{\partial}{\partial x_i}\left[\eta(T)\frac{\partial T}{\partial x_i}\right]+\frac{\rho\,\gamma_v\,(T_v-T)}{\tau_{vt}}$$

$$+\frac{\gamma(\gamma-1)\eta(T)\,\mathrm{M}^2}{2\mathrm{Re}}\left[\left(\frac{\partial u_i}{\partial x_j}+\frac{\partial u_j}{\partial x_i}\right)^2+2\left(\alpha_1-\frac{2}{3}\right)\left(\frac{\partial u_i}{\partial x_i}\right)^2\right],$$

$$\gamma_v\rho\left(\frac{\partial T_v}{\partial t}+u_j\frac{\partial T_v}{\partial x_j}\right)=\frac{\gamma\,\alpha_2}{\mathrm{Re}\,\mathrm{Pr}}\frac{\partial}{\partial x_i}\left[\eta(T)\frac{\partial T_v}{\partial x_i}\right]-\frac{\gamma_v\rho\,(T_v-T)}{\tau_{vt}},$$

$$\gamma \mathrm{M}^2 p=\rho T,\quad \eta(T)=T^{4/5},\quad i,\,j=1,\,2, \tag{7.15}$$

where the summation is assumed over the repeating indices.

Unlike the previous subsection, Eq. (7.15) involves the temperature dependence of the transport coefficients (see also Chap. 3). The power temperature dependence chosen here corresponds to the conditions of a relatively cold carrier flow (soft potentials of intermolecular interaction). At the same time the specific heats are assumed to be constant and independent of the static and vibrational flow temperatures. This enables a more complete treatment of the dissipative effect of the "blowdown" of the vibrational mode energy on the translational degrees of freedom, accompanied by an increase in gas shear viscosity.

The transport coefficients, specific heats, and the parameters involved in the equations of system (7.15) are determined as in Chap. 2. In dimensionless variables the stationary carrier flow is defined by the relations

$$U_s(x_2)=\tanh x_2,\quad T_s=T_{v,\,s}=\rho_s=1,\quad p_s=1/\left(\gamma\mathrm{M}^2\right). \tag{7.16}$$

At the time instant $t=0$ a two-dimensional disturbance with a wavelength λ and a wave vector $\mathbf{k}=(\alpha,\,0)$, where $\alpha=2\pi/\lambda$, is superimposed onto the main flow. The original boundary-value problem is posed in an infinite band whose center coincides with the origin: $x_1\in[-x_{1,\,0};\ x_{1,\,0}]$, $x_2\in(-\infty;\ \infty)$. The bandwidth along the x_1 coordinate is chosen to be equal to the disturbance wavelength $\lambda=2\pi/\alpha$ and $x_{1,\,0}=\pi/\alpha$. In calculations the asymptotic conditions at $x_2\to\pm\infty$ are translated to $x_2=\pm x_{2,\,0}$, where the ordinate $x_{2,\,0}$ is determined from the condition of the attainable computer accuracy $\left|\tanh x_{2,\,0}-1\right|\le 10^{-12}$. The value $x_{2,\,0}=20$ is finally taken. Periodic conditions are set at the computational domain boundaries at $x_1=\pm x_{1,\,0}$ at all the time, and the free-stream conditions (7.16) are imposed at $x_2=\pm x_{2,\,0}$. The initial conditions for the velocity field and thermodynamic quantities, which include the disturbances, are specified in the form

$$u_1(0,\,x_1,\,x_2)=\tanh x_2+u_1'(0,\,x_1,\,x_2),\quad u_2(0,\,x_1,\,x_2)=u_2'(0,\,x_1,\,x_2),$$

$$\rho(0,\,x_1,\,x_2)=1+\rho'(0,\,x_1,\,x_2),\quad T(0,\,x_1,\,x_2)=1+T'(0,\,x_1,\,x_2),$$

$$T_v(0,\,x_1,\,x_2)=1+\xi\,T_v'(0,\,x_1,\,x_2),$$

where the parameter ξ defines the amplitude of the vibrational temperature disturbance. It was shown in [10, 11] that the values of the parameter ξ for diatomic gases in nozzle flows, in underexpanded jets, and in flows with moderate laser pumping of vibrational modes, where dissociation can be neglected, lie in the interval $\xi = 1$–5.

The initial disturbances of the velocity vector components u_1', u_2' and thermodynamic quantities ρ', T', and T_v' introduced in the main flow are linear inviscid eigen oscillations with the highest growth rates, which were calculated in Chap. 2 (see also [12]) on the basis of a linearized system of equations of two-temperature system of aerodynamics. Table 2.1 in Chap. 2 shows the eigenvalues of the phase velocities c_i, the corresponding wave numbers α, and growth rates $\alpha\, c_i$ of the most unstable inviscid modes for the carrier flow Mach numbers M = 0–1, the vibrational relaxation time $\tau_{vt} = 1$, and the values of the parameter $\gamma_v = 0, 0.667$ of these the initial disturbances.

For computing the evolution of disturbances Eq. (7.15) are approximated by the Kovenya–Yanenko weighted finite-difference scheme [2] with splitting in terms of physical processes and spatial variables. As applied to system (7.15) this scheme was studied in Chap. 6 (see also [12]). The computations are conducted on a uniform grid with a step h in both spatial variables. A cyclic sweep is performed along the periodic x_1 coordinate; the algorithm of its implementation was taken from [4]. The steps in time and space are chosen to be equal to $h = 0.025$ and $\tau = 0.01$.

Flow regimes are determined by the following values of the parameters: Re = 100, $\mathrm{M} = 0.2, 0.5, \mathrm{Pr} = 0.75, \gamma = 1.4, \alpha_1 = 0$–$2, \xi = 1$–$5, \gamma_v = 0, 0.667$, and $\tau_{vt} = 0$–1. At these values of the parameters corresponding to the case of diatomic gases, we previously considered the dissipative effect of vibrational relaxation for a model problem in Chap. 6.

7.2.2 *Evolution of Disturbances in a Vibrationally Nonequilibrium Diatomic Gas*

The vortex disturbance evolution is traced by considering the behavior of the contours of the total vorticity of the flow ω. The grid function is calculated with use of the second-order approximation of the transverse component of the vorticity vector by central differences of the velocity field components. The computations enable to reproduce in detail the known pattern of nonlinear dynamics of a large vortex structure "cat's eye" in process of emergence and development of the Kelvin–Helmholtz instability [6]. The structure reaches its maximum size fairly quickly (at $t = 3$) and then starts to decay. By the time $t = 6$, which can be conventionally taken as the "lifetime", the structure is minimized, though its size slightly exceeds the initial value. Examples of vorticity contours in the vortex at the instant when it reaches the maximum size are presented in Fig. 7.10 for different values of the vibrational nonequilibrium degree ξ.

A comparison of the contours in Fig. 7.10 shows that, at the same initial amplitude of the velocity (vorticity) field perturbation, a greater initial disturbance of the

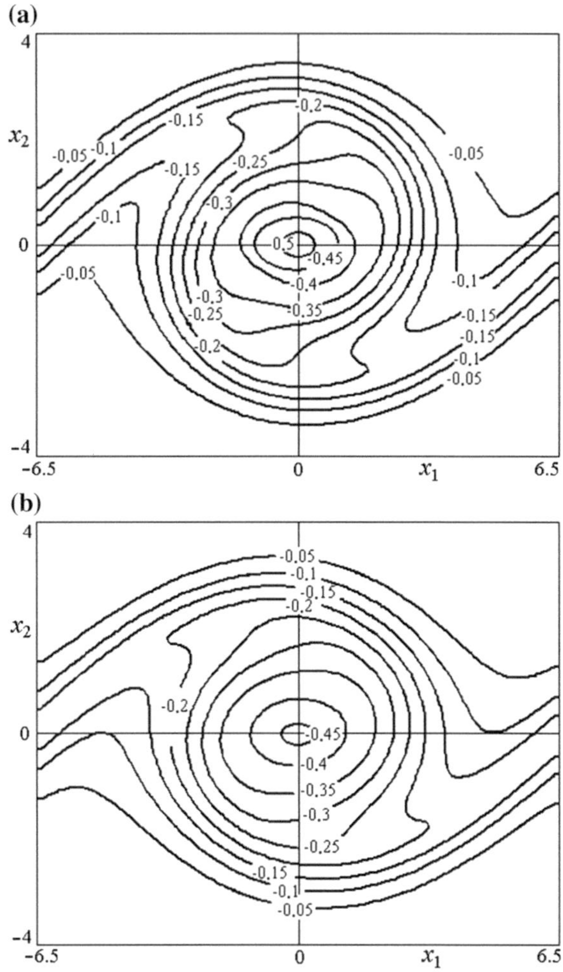

Fig. 7.10 Vorticity field contours at M = 0.5, $\alpha_1 = 0$, $\gamma_v = 0.667$, and $\tau_{vt} = 1$ at point of time $t = 3$. **a** is $\xi = 1$. **b** is $\xi = 5$

vibrational mode (see Fig. 7.10b) reduces the gradient of ω in the core of the resultant vortex structure. In the case considered this fact is explained by a higher gas temperature in the core and, correspondingly, by a high value of the shear viscosity $\eta(T)$ which leads to more intense diffusion of vorticity.

When the thermal equilibrium is approached, the energy overflow from the excited vibrational mode to the translational degrees of freedom leads to an increase in the static temperature of the gas. This result of the relaxation process is illustrated in Fig. 7.11, which shows the profiles of the vibrational and static temperatures in the vortex cross section. The temperature distributions are compared for two amplitudes of disturbances of the vibrational mode: at the initial time $t = 0$ (curves *1*) and at $t = 3$ (curves *2*) when the structure reaches its maximum size. At higher disturbance amplitude ξ of the vibrational mode, relaxation is seen to proceed more intensively.

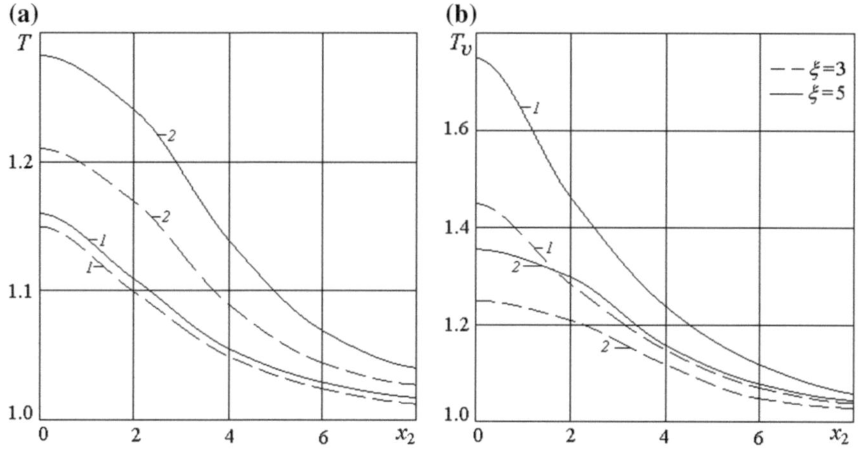

Fig. 7.11 Profiles of the static $T(x_2)$ and vibrational $T_v(x_2)$ temperatures in section $x_1 = 0$ at $\text{Re} = 100$, $\text{M} = 0.5$, $\alpha_1 = 0$, $\gamma_v = 0.667$, and $\tau_{vt} = 1$ at points of time $t = 0$ (*1*) and 3 (*2*). **a** shows profiles of the static temperatures $T(x_2)$. **b** shows profiles of the vibrational temperatures $T_v(x_2)$

The static temperature in the vortex core, which is conventionally identified by closed vorticity contours (see Fig. 7.10), turns out to be higher than at the flow periphery. This is due to the fact that the gas that entered the vortex core remains therein during the entire "lifetime" of the vortex, whereas the gas outside the core is entrained by convection out of the computational domain, and relaxation continues outside its limits.

In turn gas heating leads to an increase in gas viscosity and to a decrease in its density in the vortex core. An expected consequence should be additional decay of flow velocity fluctuations. This effect was observed in [13], where the Karman street behind a cylinder in a vibrationally excited gas was modeled. We failed to identify this channel of disturbance suppression in a clear form in our computations. A possible explanation of this fact is that the typical relaxation time, the amplitude of disturbances, and the time needed for the vortex to pass through the computational domain in [13] were substantially higher than the corresponding values used in this work.

The relaxation process unaffected by convection can be observed at the vortex center. The corresponding time evolution of the static and vibrational temperatures to thermodynamic equilibrium is shown in Fig. 7.12.

Figure 7.12a shows the relaxation curves for two values of the initial disturbance. Figure 7.12b presents the relaxation process for two values of the characteristic relaxation time τ_{vt}. As it follows from the Landau–Teller equation, the reduction of the vibrational temperature at the vortex center is determined not only by relaxation, but also by molecular heat conduction. At the same time the T_v curves, especially those in Fig. 7.12b, demonstrate mainly an exponential decrease. This circumstance agrees quite well with the estimates of the dynamics of both processes, which can be

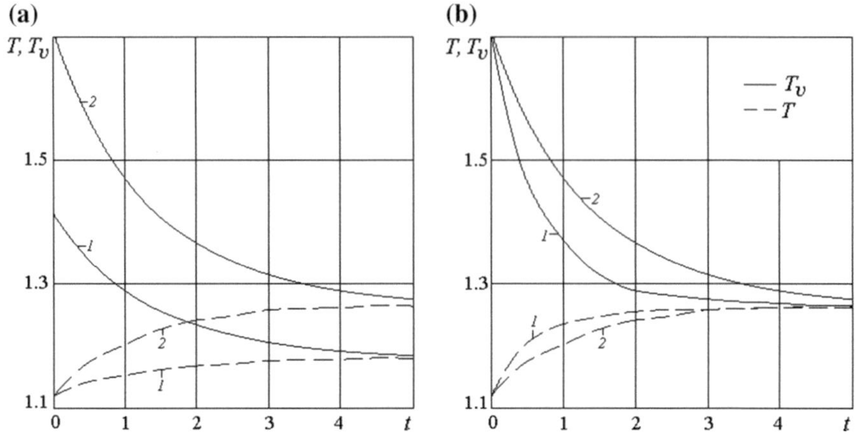

Fig. 7.12 Time evolution of the static $T(t)$ and vibrational $T_v(t)$ temperatures at the center of vortex structure at Re $= 100$, M $= 0.5$, $\alpha_1 = 0$, and $\gamma_v = 0.667$. **a** is $\tau_{vt} = 1$, $\xi = 3$ (*1*) and 5 (*2*). **b** is $\xi = 5$, $\tau_{vt} = 0.5$ (*1*) and 1 (*2*)

obtained from the equation for T_v. Indeed, the pure relaxation gives us an estimation

$$T_v(t) \sim T_v(0)\exp(-t/\tau_{vt}).$$

At the same time the heat-conduction process is estimated as

$$T_v(t) \sim \frac{\mathrm{Re\,Pr}}{t}\exp\left(-\frac{\mathrm{Re\,Pr}}{t}C\right).$$

It is seen from here that the diffusion process is substantially slower than the relaxation process at the regime parameters under consideration.

From the viewpoint of physics and possible applications it is of interest to consider the influence of vibrational relaxation on dissipation of the kinetic energy of disturbances. For this purpose time evolution of the kinetic energy of disturbances and the absolute value of Reynolds stresses are considered:

$$E(t) = \frac{1}{2}\int_{-x_{1,0}}^{x_{1,0}}\int_{-x_{2,0}}^{x_{2,0}} \rho\left(u_1'^2 + u_2'^2\right)dx_1\,dx_2, \quad \sigma_{12}(t) = \int_{-x_{1,0}}^{x_{1,0}}\int_{-x_{2,0}}^{x_{2,0}} \left|\rho u_1' u_2'\right| dx_1\,dx_2\,.$$

The equation for disturbance energy production $D(t)$ has the form [10]

$$D(t) = \frac{dE}{dt} = J_1 + J_2 - \frac{1}{\mathrm{Re}}\left(J_3 + \alpha_1 J_4\right), \tag{7.17}$$

where

$$J_1 = -\int_{-x_{1,0}}^{x_{1,0}} \int_{-x_{2,0}}^{x_{2,0}} \rho u_1' u_2' \left(\frac{dU_s}{dx_2}\right) dx_1 dx_2, \quad J_2 = \int_{-x_{1,0}}^{x_{1,0}} \int_{-x_{2,0}}^{x_{2,0}} p' \left(\frac{\partial u_1'}{\partial x_1} + \frac{\partial u_2'}{\partial x_2}\right) dx_1 dx_2,$$

and in the dissipative integrals a variable viscosity coefficient is included:

$$J_3 = \int_{-x_{1,0}}^{x_{1,0}} \int_{-x_{2,0}}^{x_{2,0}} \eta(T) \left[\left(\frac{\partial u_1'}{\partial x_1}\right)^2 + \left(\frac{\partial u_1'}{\partial x_2}\right)^2 + \left(\frac{\partial u_2'}{\partial x_1}\right)^2 + \left(\frac{\partial u_2'}{\partial x_2}\right)^2\right.$$

$$\left. + \frac{1}{3}\left(\frac{\partial u_1'}{\partial x_1} + \frac{\partial u_2'}{\partial x_2}\right)^2\right] dx_1\, dx_2, \quad J_4 = \int_{-x_{1,0}}^{x_{1,0}} \int_{-x_{2,0}}^{x_{2,0}} \eta(T) \left(\frac{\partial u_1'}{\partial x_1} + \frac{\partial u_2'}{\partial x_2}\right)^2 dx_1\, dx_2.$$

Here u_1', u_2', and p' are the fluctuations of the velocity components and pressure, $x_{1,0} = \pi/\alpha$, and $x_{2,0} = 20$. The wavenumbers α for different values of regime parameters γ_v, M, and τ_{vt} are taken from Table 7.1.

The fluctuation characteristics of the flow are determined as

$$\mathbf{F}'(x_1, x_2, t) = \mathbf{F}(x_1, x_2, t) - \mathbf{F}_s(x_1, x_2, t),$$

where the components of the vector function **F** are the instantaneous values of the disturbed flow characteristics, and $\mathbf{F}_s$ are the corresponding characteristics of the carrier flow. As the undisturbed flow (7.16) is not an exact stationary solution of system (7.15) in the case considered, its instantaneous characteristics are calculated in parallel with the disturbed flow computation. The integrals in Eq. (7.17) are computed by quadrature trapezoid formulas with a step $h = 0.025$ on the grid used in the computations.

Examples of the time evolution of the kinetic energy of disturbances $E(t, \xi)$ and the fluctuation energy production $D(t, \xi)$ for some regimes are presented, respectively, in Figs. 7.13 and 7.14. It is seen that stratification of the $E(t)$ curves in terms of the vibrational energy nonequilibrium parameter ξ is sufficiently appreciable, and the evolution of the structure for all values of ξ has a universal character: growth and reaching the maximum value of the structure energy at $t = 3$, then its decrease and stabilization at $t > 5$ to a value slightly exceeding the initial one.

As follows from the graphs in Fig. 7.13, a higher value of the parameter ξ corresponds to a lower kinetic energy of the structure in the entire time interval. An increase in the depth of excitation of the vibrational mode ξ, as well as an increase in the bulk viscosity η_b (parameter α_1), leads to enhanced dissipation of the kinetic energy of disturbances. The absolute values of Reynolds stresses as a function of time $|\sigma_{12}|(t, \xi, \alpha_1)$ practically repeat the graphs of $E(t, \xi, \alpha_1)$. At the same time the computations show that the variation of the vibrational relaxation time τ_{vt} in the limits 0.1–1 exerts a minor effect on the behavior of $E(t)$ and $|\sigma_{12}|(t)$.

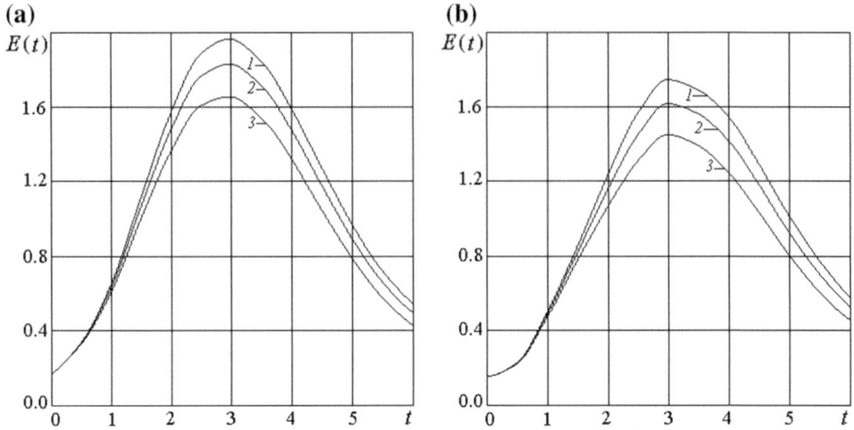

Fig. 7.13 Time evolution of the disturbance energy $E(t)$ at Re $= 100$, M $= 0.5$, $\gamma_v = 0.667$, $\tau_{vt} = 1$, $\xi = 1$ (*1*), 3 (*2*), and 5 (*3*). **a** is $\alpha_1 = 0$. **b** is $\alpha_1 = 2$

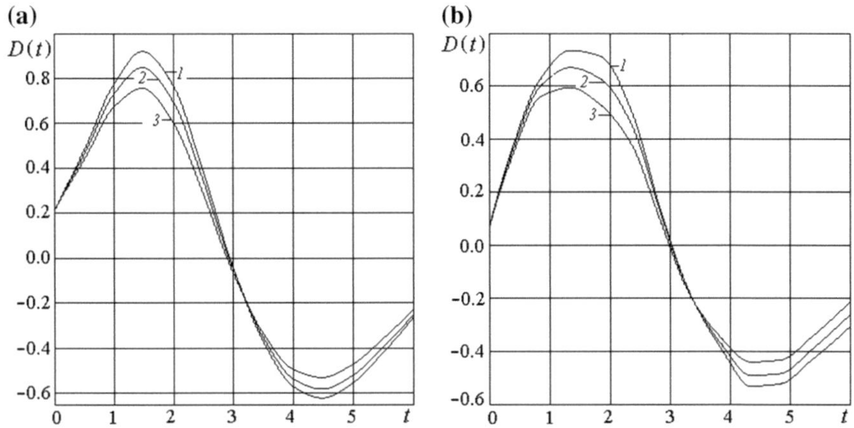

Fig. 7.14 Time evolution of the fluctuation energy production $D(t)$ at Re $= 100$, M $= 0.5$, $\gamma_v = 0, 667$, $\tau_{vt} = 1$, $\xi = 1$ (*1*), 3 (*2*), and 5 (*3*). **a** is $\alpha_1 = 0$. **b** is $\alpha_1 = 2$

Table 7.6 Relative deviations $\varepsilon_E(\xi)$, % for Re $= 100$, $\tau_{vt} = 1$, and $\gamma_v = 0.667$

M	$\alpha_1 = 0$			$\alpha_1 = 1$			$\alpha_1 = 2$		
	$\xi = 1$	$\xi = 3$	$\xi = 5$	$\xi = 1$	$\xi = 3$	$\xi = 5$	$\xi = 1$	$\xi = 3$	$\xi = 5$
0.2	1.986	3.324	9.721	2.121	3.690	10.448	2.332	4.087	11.455
0.5	2.750	5.424	11.348	3.799	5.730	12.009	3.096	6.177	14.567

As is seen from a comparison of the graphs in Figs. 7.13 and 7.14, in the time interval where the kinetic energy of disturbances grows, the energy production is positive at first and increases, reaching a peak value; then it starts to decrease and becomes negative, which corresponds to a decrease in energy of the structure. The points where the $D(t)$ curves pass through zero correspond to the maxima on the $E(t)$ curves, and the maximum and minimum points of the $D(t)$ graphs respond to the inflection points on the graphs of $E(t)$.

For a quantitative comparison of the contribution of vibrational relaxation to dissipation of the kinetic energy of disturbances the relative deviations were found as

$$\varepsilon_E(\xi) = \left| \frac{\langle E(t,\,\xi)\rangle - \langle E(t,\,0)\rangle}{\langle E(t,\,0)\rangle} \right| 100\%, \quad \xi = 1 - 5,$$

where the angular brackets $< \dots >$ mean averaging over the conventional "lifetime" of the structure $\Theta = 6$. The computed relative deviations $\varepsilon_E(\xi)$ for some combinations of parameters are presented in Table 7.6.

It follows from Table 7.6 that an increase in the parameter ξ at fixed values of the Mach number M and parameters α_1, τ_{vt} leads to more intense dissipation of the kinetic energy of disturbances. In particular, in the absence of bulk viscosity ($\alpha_1 = 0$), the relative decrease in the time-averaged energy $\langle E(t,\,\xi)\rangle$ reaches $\varepsilon_E = 11.35\%$ for $\xi = 5$ and $\mathrm{M} = 0.5$. This value agrees well with the results of Chap. 6 (see also [14]) for dissipation of the disturbance energy $\langle E(t,\,\xi)\rangle$ only due to relaxation of the vibrational mode of molecules in the absence of other dissipative processes in the flow. Although the model problem considered in Chap. 6 [14] reproduced only the structure decay which corresponds in our case to the descending branches of the curves in Fig. 7.13. The mean relative suppression of the kinetic energy of disturbances amounts in our computations to $\varepsilon_E = 13.10\%$ for the maximum values of bulk viscosity at $\alpha_1 = 2$, Mach number $\mathrm{M} = 0.5$, and $\xi = 5$.

It can be noted that Table 7.6 illustrates a unidirectional character of the effects of compressibility, bulk viscosity, and excitation of vibrational degrees of freedom: enhancement of these effects leads to more intense dissipation of disturbance energy. At the same time it is seen that admissible excitations of the vibrational modes ensure a much greater higher dissipation effect as compared to the possible contribution of bulk viscosity. The relative fraction of the latter does not exceed 15%. This allows one to consider forced pumping of the vibrational modes as a real tool of action on disturbances in molecular gas flows.

Numerical modeling results of nonlinear development of subsonic vortex disturbances in a shear layer of a vibrationally excited gas, which evolves in time, confirm once again the conclusion made in previous chapters about an appreciable damping influence of the relaxation of excited internal molecular modes on the linear and nonlinear dynamics of disturbances. Let us note that the range of flow parameters, in particular, the ratios of the coefficients of the bulk and shear viscosities and the non-equilibrium parameter of the vibrational mode corresponds to the actually attainable values for diatomic gases.

The range of variation of the dissipative effect during vibrational mode relaxation is comparable with the mechanical technique of large eddy breakup (LEBU) [15] or ribletting surfaces. Therefore, laser pumping of vibrational modes may become a new realistic technique of molecular gas flow control.

References

1. Blumen, W.: Shear layer instability of an inviscid compressible fluid. J. Fluid Mech. **40**, 769–781 (1970)
2. Kovenya, V.M., Yanenko, N.N.: Splitting Method in Problems of Gas Dynamics. Nauka, Novosibirsk (1981) (in Russian)
3. Grigor'ev, Yu.N, Ershov, I.V., Zyryanov, K.V., Sinyaya, A.V.: Numerical simulation of the bulk viscosity effect on a sequence of nested grids. Vych. Tekhnol. **11**, 36–49 (2006) (in Russian)
4. Kvasov, B.I.: Interpolation by Cubic and Bicubic Splines. Novosibirsk State University Publication, Novosibirsk (2004) (in Russian)
5. Betchov, R., Criminale, W.O.: Stability of Parallel Flows. Academic Press, New York (1967)
6. Patnaik, P.C., Sherman, F.S., Corcos, G.M.: A numerical simulation of Kelvin–Helmholtz waves of finite amplitude. J. Fluid Mech. **73**, 215–239 (1976)
7. Corcos, G.M., Sherman, F.S.: The mixing layer deterministic models of a turbulent flow. Part I. Introduction and two-dimensional flow. J. Fluid Mech. **139**, 29–65 (1984)
8. Stepanov, V.V.: Course of Differential Equations. Fizmatlit, Moscow (1959) (in Russian)
9. Grigor'ev, Yu.N, Ershov, I.V.: Relaxation-induced suppression of vortex disturbances in a molecular gas. J. Appl. Mech. Tech. Phys. **44**, 471–481 (2003)
10. Grigor'ev, Yu.N, Ershov, I.V., Zyryanov, K.V.: Numerical modeling of Kelvin–Helmholtz waves in a weakly nonequilibrium molecular gas. Vych. Tekhnol. **13**, 25–40 (2008) (in Russian)
11. Osipov, A.I., Uvarov, A.V.: Kinetic and gas-dynamic processes in nonequilibrium molecular physics. Usp. Fiz. Nauk **162**, 1–42 (1992) (in Russian)
12. Grigor'ev, Yu.N, Ershov, I.V., Zyryanov, K.V.: Numerical modeling of inertial instability in a vibrationally nonequilibrium diatomic gas. Vych. Tekhnol. **15**, 42–57 (2010) (in Russian)
13. Vinnichenko, N.A., Nikitin, N.V., Uvarov, A.V.: Karman vortex street in a vibrationally nonequilibrium gas. Fluid Dyn. **40**, 762–768 (2005)
14. Grigor'ev, Yu.N, Ershov, I.V., Ershova, E.E.: Influence of vibrational relaxation on the pulsation activity in flows of an excited diatomic gas. J. Appl. Mech. Tech. Phys. **45**, 321–327 (2004)
15. Savill, A.M.: Drag reduction by passive devices - a review of some recent developments. In: Gyr, A. (ed.) Structure of Turbulence and Drag Reduction, pp. 429–465. Springer, Berlin (1990)

Index

Y.N. Grigoryev and I.V. Ershov, *Stability and Suppression of Turbulence in Relaxing Molecular Gas Flows*, Fluid Mechanics and Its Applications 117,
DOI 10.1007/978-3-319-55360-3

MIX
Papier aus verantwortungsvollen Quellen
Paper from responsible sources
FSC® C105338
www.fsc.org

If you have any concerns about our products,
you can contact us on
ProductSafety@springernature.com

In case Publisher is established outside the EU,
the EU authorized representative is:
Springer Nature Customer Service Center GmbH
Europaplatz 3, 69115 Heidelberg, Germany

Printed by Libri Plureos GmbH
in Hamburg, Germany